Daiane Corrêa
Suelen Cristina Uber

ALLELOPATHY IN AGRICULTURE

Daiane Corrêa
Suelen Cristina Uber

ALLELOPATHY IN AGRICULTURE

Action on vegetable plants

ScienciaScripts

Imprint
Any brand names and product names mentioned in this book are subject to trademark, brand or patent protection and are trademarks or registered trademarks of their respective holders. The use of brand names, product names, common names, trade names, product descriptions etc. even without a particular marking in this work is in no way to be construed to mean that such names may be regarded as unrestricted in respect of trademark and brand protection legislation and could thus be used by anyone.

Cover image: www.ingimage.com

This book is a translation from the original published under ISBN 978-620-2-80873-6.

Publisher:
Sciencia Scripts
is a trademark of
International Book Market Service Ltd., member of OmniScriptum Publishing Group
17 Meldrum Street, Beau Bassin 71504, Mauritius
Printed at: see last page
ISBN: 978-620-3-38229-7

ALLELOPATHY IN AGRICULTURE

Action on vegetable plants

DAIANE CORRÊA
SUELEN CRISTINA UBER

SUMMARY

Chapter 1

Allelopathy in agriculture and its effects on plants

Daiane Corrêa

Allelopathy in Agriculture

Allelopathy, from the Latin *allelon* = mutual, and *pathos* = harm, is defined as any substance that comes from plants or living organisms, which, when released into the environment tend to cause problems to the surrounding community or cause beneficial effects (BARATELLI, 2006).

Allelopathy is a phenomenon with different interactions that can be found among all organisms, however, it is in plants that it is more noticeable to be verified. This phenomenon is a strategy acquired over the years by plants, as a form of protection against the attack of pathogens, insect pests and herbivory (ALMEIDA, 1991).

Differently from the primary metabolite, whose compounds are amino acids, nucleotides, lipids, carbohydrates and chlorophyll, and present a wide distribution in the plant. The compounds of the secondary metabolite are not distributed throughout the plant, presenting different concentrations in different parts of the plant tissues (ALVES, 2009), originating from the shikimate or acetate route, or even their combined arrangement (DIAS et al., 2005).

Almeida (1991) also points out that, even if the plants are dead, their tissues have the ability to release allelochemical substances, both by volatilization, and by leaching substances from their tissues, which are soluble in rainwater or dew, and are conducted to the soil, causing interference in nearby microorganisms and plants.

Several times the occurrence of allelopathy can be confused with the effects caused by competition between plants, due to the influence of both on growth and development of plants (SILVA, 2012). However, the two are different, since competition is the action of plants interfering for the absorption of other resources (RIZZARDI et al., 2001), such as essential nutrients for growth and development, light, water and space (PITELLI, 1987), in the case of allelopathy, there is no dispute for the need for resources, but there is an introduction of allelochemical compounds into the environment, causing primary and secondary effects on the other plant (SANTOS, 2007).

The action of allelochemicals may occur directly and indirectly (FERREIRA; AQUILA, 2000). According to Borella et al. (2010) the direct action consists of intervention in the development and plant metabolism, including modifications at the cellular, phytormonal, photosynthetic and respiratory levels, impairs DNA and RNA synthesis, protein synthesis or inhibits and stimulates specific enzyme activities.

Ferreira and Aquila (2000) complement that, in direct action, the allelochemicals tend to bind in plant membranes, penetrating into their cells and successively intervening in their metabolism. In the case of indirect action, there is a change in soil properties and in its nutritional condition, affecting the populations or functions of microorganisms (FERREIRA; AQUILA, 2000).

In today's agriculture, the term "soil depletion" has been frequently described, a fact that is directly linked to aspects that cause the reduction of fertility, as well as competition between species, however, several studies prove that in addition to these factors, allelopathy is intrinsically correlated to this problem (GATTI, 2008), because such substances derived from allelopathy between plants, present in the soil, interfere with seed germination, growth and development of plants, thus preventing their establishment in agricultural crop areas (CORSATO et al., 2010).

Allelochemical Applications

The first studies on allelopathy were restricted mainly to European countries and North America, over the years this science has spread to countries in South America, Asia and the Middle East, thus promoting the aggregation of new experimental methods and isolation of allelochemical substances present in plants (SOUZA FILHO; GUILHON; SANTOS, 2010).

The identified substances exceed the mark of 10,000 thousand (EMBRAPA, 2001), among them Dias et al. (2005) highlight the main classes subdivided into compounds derived from terpenes, alkaloids, phenolics, steroids, long-chain fatty acids and unsaturated lactones. However according to Santos et al. (2011), it is the terpenoid and phenolic classes that represent the majority of compounds that cause allelopathy in the field.

With the advances in research, chromatography techniques and the discovery of new allelochemical substances with different structural forms, it was observed the great potential in the making of cosmetics, perfumery and even in the health area (SANTOS, 2015). In addition, it has become a great strategy in agriculture as nematicide, insecticide (FERREIRA; AQUILA, 2000) and herbicide, reducing the application of pesticides, and reducing or even eliminating the contamination of the environment (FORMAGIO et al., 2010).

Another way of applying the allelochemicals is through the use of cover plants, which, by releasing water-soluble compounds during their decomposition, become an alternative and/or strategy to limit the growth of weeds, for example in beet plantings or in vegetables (MARTINELLI; SILVA, 2018).

Positive and negative effects of allelopathy

The influence of one plant on the development of others that are nearby has been observed since antiquity (REZENDE et al., 2003), and it is not caused simply by an isolated factor, but by several substances acting together with the action of the environment (COMIOTTO, 2006).

In many cases it can be observed the absence of plants of the same species under the canopy of some fruit trees, as in the case of mango, and this is called by Comiotto (2006) as "autotoxicity", i.e. the release of allelochemical substances that prevent the germination and development of plants of the same species. The same author points out that when this allelochemical action occurs in different plant species it is called "heterotoxicity".

Heterotoxicity, so named by Comiotto (2006), is a negative effect of allelopathy, and is observed in vegetables near areas of cultivation with *eucalyptus (Eucalyptus* sp.), which suffered with the inhibition of seed germination and seedling development (CREMONEZ et al., 2013). Being observed the same effect of neen (*Azadirachta indica* A.), which has allelopathic substances such as azadirachtin, causing interference in the germination of sorghum and lettuce seeds (FRANCE et al., 2008). Silva (2017) writes that it was possible to verify negative effects of extracts of cidergrass and milfoil flowers on the germination of vegetable seeds, such as lettuce and cucumber.

Thyrrhiza (*Cyperus rotundus* L.) is another classic example of a plant that releases allelopathic substances, causing negative effects. It is considered one of the main weeds in several agricultural crops, in crops such as sugarcane, it interferes during its tillering, leading to a smaller number of plants per cultivated area (PASTRE, 2006). This interference also occurs during the germination of seeds of corn and other poaceae, as identified by Scheren; Ribeiro and Nobrega (2014). Other studies show that extracts of different species of Annonaceae, affected the germination and development of lettuce (FORMAGIO et al., 2010).

Among the positive effects of allelochemicals evidenced in the literature, it can be highlighted the presence of indole acetic acid (IAA) and indole butyric acid (IAB) in leaves and roots of tiririca (*C. rotundus* L.), which are rooting inducers (NAVARRO; SILVA;

MOECKE, 2017). Cremonez et al. (2013) point out that, the capim-santo (*Cymbopogon citratus*) stimulates the development of other plants. Alves et al. (2004) observed that the volatile extract of jamborandi (*Pilocarpus microphyllus* Stapf. ex. Wardleworth) promotes rooting in lettuce, thus evidencing that allelopathic substances can act in a positive or negative way, through the joint or isolated actions of their compounds of secondary metabolism.

Cyrillic (*Cyperus rotundus* L.)

Thyririca is a difficult to control weed that causes damage in a wide range of crops of commercial importance due to its competition throughout the cycle of agricultural crops, and the initial period after the implementation of the crop in the field is the most critical in which this competition occurs more severely (SILVEIRA et al., 2010).

The center of origin and diffusion of tiririca was in Southeast Asia, mainly in India. The weed is cosmopolitan and can be found in countries with tropical, subtropical and temperate climates (PASTRE, 2006). According to Pastre (2006), the possible installation of tiririca in Brazilian territory occurred through the merchant ships, from Portugal, with the European colonizers, where it was first identified in the port cities, such as Salvador, Recife, Rio de Janeiro, Santos and São Vicente, where they established and multiplied and later spread to the interior of the country.

Belonging to the Cyperaceae family, tiririca is similar to the grasses, differing due to its form of propagation, which can be by seeds, however they have low viability, with germination of approximately 5%, and its most efficient way of multiplication is through basal bulbs, stoloniferous rhizomes and tubers, making its control more difficult because they are underground multiplication systems (MIRÓ; FERREIRA; AQUILA, 1998).

The tiririca is an herbaceous plant that performs photosynthesis through the C4 cycle, very efficient in hot climates, and in Brazilian climatic conditions it reaches an average size of 15-30 cm, considered an invasive plant with great capacity for survival in adverse conditions (PASTRE, 2006).

The diversity of soil and climatic conditions in each region means that tiririca has a different life cycle. Although it is considered a perennial plant, in conditions of low soil humidity, its metabolic activities are reduced and the tubers go dormant until they find suitable conditions in the soil to resume their development. Tubers can have a longevity of between 3 and 5 years, which can be longer depending on the depth at which they are found (MIRÓ; FERREIRA; AQUILA, 1998).

In agro-ecological cultivation, the management of population reduction and control of tiririca are difficult to achieve, and may interfere with the development of several crops of agricultural importance. Among the alternatives, to reduce the damage caused by weeds, is the adoption of green manure in conjunction with conservationist practices, in which the plants used in this model tend to form a barrier on the soil surface, preventing the twinning of the weed seed bank, as well as the suppression of seedlings due to competition for light, water and nutrients. In this context, allelopathy can be used as a way to minimize the development of weeds, through the exudation of allelochemicals derived from secondary metabolism, without interfering in agricultural crops of economic interest (FONTANETTI et al., 2007).

Clove (*Tagetes* spp.)

The clove, known as *Tagetes L.*, belonging to the family Asteraceae, contains approximately fifty species, annual or perennial, with *T. minuta, T. erecta, T. patula and T. tenuifolia being* the most commonly found (VASUDEVAN et al., 1997).

Although *Tagetes* is native to Mexico and other warm regions of America, it is cultivated throughout the world as an ornamental plant. It is usually presented as shrubs, with the presence in its leaves and bracts, of glands containing essential oils (BANO et al., 2002).

According to Xu et al. (2012), 126 substances present in species of the genus *Tagetes* were identified, such as phenolic compounds, alkaloids, carotenoids, thiophenes, benzofurans, terpenoids, steroids, and coumarins, with the predominance of some of these in specific organs of the plant. As an example, the essential oil and flavonoids are preferentially produced in the leaves and flowers; the carotenoids are only present in the petals and the thiophenes are mainly found in the roots (MAROTTI et al., 2010; MUNHOZ, 2013).

Clove has medicinal and therapeutic properties, which have been recognized since the time of the Aztecs, as well as antimicrobial, biological activities such as insecticide, larvicide and bactericide (SIRI HARTATI et al., 1999).

It can also be used in phytonematoid control, especially against *Pratylenchus* and *Meloidogyne* species, as well as in the manufacture of natural dye for food and beverages, poultry feed supplement (PADMA et al., 1997).

References

ALENCAR, G. Unpublished study identifies major weeds to tomatoes intended for industry. **EMBRAPA**, 2018. Available at: https://www.embrapa.br/ busca-de-noticias/-/noticia/34802765/estudo-inedito-identifica-principais-plantas-daninhas-ao-tomate-destinado-a-industria. Accessed on: 14 Mar. 2020.

ALMEIDA, F. S. Allelopathic effects of plant residues. **Revista Pesquisa Agropecuária Brasileira,** Brasília, v. 10, p. 221-236, 1991.

ALVES, J. N. Chemical Characterization of Dichloromethane Extracts of *Origanum majorana* L. in the Inhibition of *Panicum maximum*. 2009. 96 f. Master (Post Graduation Chemistry) - Universidade Federal de Uberlândia - UFU, Uberlândia. 2009.

ALVES, M. C. S.; FILHO, S. M.; INNECCO, R.; TORRE, S. B. Allelopathy of volatile extracts on seed germination and root length of lettuce. **Pesquisa Agropecuária Brasileira,** Brasília, v. 39, n. 11, p. 1083-1086, 2004.

AMARAL, M. C. A.; BANDEIRA, A. S.; PORTO, J. S.; ÁVILA, J. S.; SANTOS, R. K. A.; MORAIS, O. M. Avaliação do efeito allelopático de extrato aquoso de tiririca sobre a germinação de sementes de carrot. **Cadernos de Agroecologia,** Seropédica, v. 13, n. 1, p. 1-7, 2018.

ANDRADE, H. M.; BITTENCOURT, A. H. C.; VESTENA, S. Allelopathic potential of Cyperus rotundus L. on cultivated species. **Ciência e Agrotecnologia,** Lavras, v. 33, n. SPE, p. 1984-1990, 2009.

BANO, H. et al. Chemical constituents of *Tagetes patula* L. **Journal Pharm Science**, v. 15, n. 2, p. 1-12, 2002.

BARATELLI, T. G. **Study of plant allelopathic properties: investigation of allelochemical substances in Terminalia catappa L. (Combretaceae).** 2006. 206 f. Thesis (Doctorate in Natural Products Chemistry) - Universidade Federal do Rio de Janeiro - UFRJ, Rio de Janeiro. 2006.

BORELLA, J.; PASTORINI, L. H. Allelopathic influence of Phytolacca dioica L. on germination and initial growth of tomato and black pepper. **Revista Biotemas**, Florianópolis, v. 22, n. 3, p. 67-75, 2009.

BORELLA, J.; TUR, C. M.; PASTORINI, L. H. Allelopathy of aqueous extracts of *Duranta repens* on germination and initial growth of Lactuca sativa and Lycopersicum esculentum. **Revista Biotemas**, Florianópolis, v. 23, n. 2, p. 13-22, 2010.

BRASIL. Ministry of Agriculture, Livestock and Supply. **Regras para análises de sementes**. Brasília: MAPA/ACS, 2009. 398p.

BRZEZINSKI, C. R.; ABATI, J.; GELLER, J.; WERNER, F.; ZUCARELI, C. Production of American lettuce cultivars under two cropping systems. **Revista Ceres**, Viçosa, v. 64, n. 1, 2017.

CASTRO, P. R. C.; RODRIGUES, J. D.; MORAES, M. A.; CARVALHO, V. L. M. Allelopathic effects of some plant extracts on tomato (*Lycopersicon esculentum* Mill. cv. Santa Cruz) germination. **Planta Daninha**, Piracicaba, v. 6, n. 2, p. 79-85, 1983.

CAVALCANTE, J. A.; LOPES, K. P.; PEREIRA, N. A. E.; SILVA, J. G.; PINHEIRO, R. M.; MARQUES, R. L. L. Aqueous extract of thyrrhiza bulbs on germination and initial growth of radish seedlings. **Revista Verde de Agroecologia e Desenvolvimento Sustentável**, Pelotas, v. 13, n. 1, p. 39-44, 2018.

COMIOTTO, A. **Allelopathic potential of different plant species on the physiological quality of rice seeds and lettuce achenes and growth of rice and lettuce seedlings.** 2006. 42 f. Dissertation (Master of Science) Universidade Federal de Pelotas - UPF, Pelotas. 2006.

CONAB. **Boletim hortigranjeiro**. Brasília. v. 5, n. 4, April 2019.

CORSATO, J. M.; FORTES, A. M. T.; SANTORUM, M.; LESZCZYNSKI, R. Allelopathic effect of aqueous extract of sunflower leaves on the germination of soybean and black walnut. **Semina: Agrarian Sciences**, Londrina, v. 31, n. 2, p. 353-360, 2010.

CREMONEZ, F. E.; CREMONEZ, P. A.; CAMARGO, M. P.; FEIDEN, A. Principal plants with allelopathic potential found in Brazilian agricultural systems. **Acta Iguazu**, Cascavel, v. 2, n. 5, p. 70-88, 2013.

DEOMEDESSE, C. C.; MENESES, N. B.; SOUSA, G. O.; SILVA, T. S.; CRUZ, G. A. Allelopathic effects of tiririca extract on the germination of sweet corn, lettuce, cucumber, and string. **MAGISTRA**, Cruz das Almas, v. 30, p. 323-330, 2019.

DIAS, J. F.G. **Applied allelopathic study of *Aster lanceolatus*, Willd.** 2005. 46f. Dissertation (Master in Pharmaceutical Sciences). Universidade Federal do Paraná, Curitiba. 2005.

DIAS, J. F. G.; CÍRIO, G. M.; MIGUEL, M. D.; MIGUEL, O. G. Contribuição ao estudo alelopático de Maytenus ilicifolia Mart. ex Reiss., Celastraceae. **Revista Brasileira de Farmacognosia**, João Pessoa, v. 15, n. 3, p. 220-223, 2005.

DUSI, A. N. A cultura do tomatoiro (para mesa). **EMBRAPA-SPI. Coleção plantar**, 1993.

Allelopathic Studies Related to Coffee. Porto Velho: EMBRAPA, n. 54, p. 26. 2001. *ISSN 0103-9865.*

FERREIRA, A. G.; AQUILA, M. E. A. Allelopathy: an emerging area of ecophysiology. **Revista Brasileira de Fisiologia Vegetal**, Brasília v. 12, n. 1, p. 175-204, 2000.

FILHO, R. F.; SANTOS, B. V. B.; LOPES, T. B.; SOUZA, B. N.; SOUZA, B. R. Utilizacao de extrato de folhas de Tirirca in *vitro* na cultura da Alface. **In:** Congresso Nacional de Meio Ambiente, 15º. Poços de Caldas. p. 1-5, 2018.

FORMAGIO, A.; MASETTO, T. E.; BALDIVIA, D. S.; VIEIRA, M. C.; ZÁRATE, N. A. H.; PEREIRA, Z. V. Potencial alelopático de cinco espécies da família Annonaceae. **Revista Brasileira de Biociências**, Porto Alegre, v. 8, n. 4, 2010.

FONTANÉTTI, A.; CARVALHO, G. J.; GOMES, L. A. A.; DE ALMEIDA, K.; DE MORAES, S. R. G.; DUARTE, W. F. Efeito alelopático da adubação verde no controle de tiririca (Cyperus rotundus L.). **Cadernos de Agroecologia**, Seopédica, v. 2, n. 1, 2007.

FRANÇA, A. C.; SOUZA, I. F.; SANTOS, C. C.; OLIVEIRA, E. Q.; MARTINOTTO, C. Allelopathic activities of neem on the growth of sorghum, lettuce and black walnut. **Revista Ciência e Agrotecnologia**, Lavras, v. 32, n. 5, p. 1374-1379, 2008.

FUMAGALI, E.; GONÇALVES, R. A. C.; MACHADO, M. F. P. S.; VIDOTI, G. J.; OLIVEIRA, A. J. B. Production of secondary metabolites in plant cell and tissue culture: The example of the *Tabernaemontana* and *Aspidosperma* genera. **Revista Brasileira de Farmacognosia**, João Pessoa, v. 18, n. 4, p. 627-641, 2008.
GATTI, A. B. **Allelopathic activities of cerrado species**. 2008. 138 f. Thesis (PhD in Ecology and Natural Resources) Universidade Federal de São Carlos - UFSCar, SP. 2008.

GUSMAN, G. S.; YAMAGUSHI, M. Q.; VESTENA, S. Allelopathic potential of aqueous extracts of *Bidens pilosa* L., *Cyperus rotundus* L. and *Euphorbia heterophylla* L.. **Iheringia. Botany Series,** Porto Alegre, v. 66, n. 1, p. 87-98, 2011.

HADAS, A. Water uptake and germination of leguminous seeds under changing external water potencial in osmotic solution. **Journal Express Botany**, Saint Louis, v. 27, p. 480-489, 1976.

KHATOUNIAN, C. A.; OLIVEIRA, D. A. M.; FERREIRA, T. M.; DUPRE, M.; MERIANNE, H. Distribution of tubers of Tiririca (*Cyperus rotundus* L.) in the soil profile and its implications for conversion to organic agriculture of urban gardens. **Scientia Plena**, Piracicaba, v. 14, n. 9, 2018.

LAMEGO, K. B.; DA SILVA, J. Allelopathic potential of the aqueous extract of the tubers of tiririca. **South American**, Ji-Paraná, v. 4, n. 2, p. 84-93, 2017.
MAROTTI, I. et al. 2010. Thiophene ocurrence in different *Tagetes* species: agricultural biomasses as sources of biocidal substances. **Journal Science Food Agriculture,** v. 90, p. 1210-1217.

MARTINELLI, V. A.; SILVA, V. N. Allelopathic effect of rye on germination and growth of beet seedlings. **Agrarian Academy Magazine**, Goiânia, v. 5, n. 9; p. 1-9, 2018.

MUNHOZ, V. M. **Pharmacognostic evaluation and optimization of flavonoid extraction from *Tagetes patula* flowers through mixture planning**. 2013. 57 f. Dissertation (Master Pharmaceutical Sciences), Universidade Estadual de Maringa, Maringa, 2013.

MIRÓ, Cíntia Pilotti; FERREIRA, Alfredo Gui; AQUILA, Maria Estefânia Alves. Allelopathy of yerba mate (*Ilex paraguariensis*) fruits on corn development. **Pesquisa Agropecuária Brasileira**, Brasilia, v. 33, n. 8, p. 1261-1270, 1998.

MOGHARBEL, A. D. I.; MASSON, M. L. Perigos associados ao consumo da alface (*Lactuca sativa*), in natura. **Alimentação Nutritiva**, Araraquara, v. 16, n. 1, p. 83-88, 2005.

MOLISCH, H. **The influence of one plant on another.** Published by Gustav Fischer, 1937.

MORALES-PAYAN, J. P.; STALL, W. M.; SHILLING, D. G.; CHARUDATTAN, R.; DUSKY, J. A.; BEWICK, T. A. Above-and belowground interference of purple and yellow nutsedge (*Cyperus* spp.) with tomato. **Weed science**, Amsterda, v. 51, n. 2, p. 181-185, 2003.
NAVARRO, L. F. F.; SILVA, M. S.; MOECKE, U. F. R. **Efficiency of the organic extract of "*Cyperus rotundus* as a rooter in the propagation of *Corymbia citriodora*".** 2017. 55 f. Monograph (Bachelor in Agronomy) Centro Universitário Católico Salesiano *Auxilium* - UniSALESIANO, Lins. 2017.

ORTIZ, T. A.; MORITA, D. A. S.; COSMO, D. A.; SELEGUINI, A.; ARIERA, C. D.; HORA, R. C. 2008. Germination of lettuce seeds as a function of concentrations of aqueous extract of tiririca. **Revista Horticultura Brasileira**, Umuarama, v. 26, n. 2, p. 1-5, 2008.

PADMA, V. et al. *Tagetes*: a multipurpose plant. **Bioresource Technology**, v. 62, n. 1-2, p. 29-35, 1997.

PASTRE, W. **Control of thyrrhiza (*cyperus rotundus* L.) with sulfentrazone and flazasulfuron applied alone and in mixture in sugarcane crop**. 2006. 66 f. Dissertation (Master in Tropical and Subtropical Agriculture) Agronomic Institute - IAC, Campinas. 2006.

PITELLI, R. A. Competição e controle das plantas weeds em áreas agrícolas. **Série técnica IPEF**, Piracicaba, v. 4, n. 12, p. 1-24, 1987.

QUADROS, B. R.; TANAKA, A. A.; CARDOSO, A. I. I.; RIGOTTI, M.; SANTOS, R. F. Allelopathy of plant extracts of pelargonium graveolens and cyperus rotundus on lettuce seed germination. **Revista Horticula Brasileira**, Botucatu, v. 27, n. 2, 2007.

QUEIROZ, A. A.; CRUVINEL, V. B.; FIGUEIREDO, K. Production of American lettuce as a function of fertilization with organomineral. **Enciclopédia Biosfera, Centro Científico Conhecer,** Goiânia, v. 14, n. 25, p. 1-12, 2017.

REGHIN, M. Y.; PURÍSSIMO, C.; DALLA PRIA, M.; FELTRIM, A. L.; FOLTRAM, M. A. Técnicas de cobertura do solo e de proteção de plantas no cultivo da alface. **Horticultura Brasileira**, Ponta Grossa, v. 20, n. 2, p. 1-10, 2002.

REZENDE, C. de P.; PINTO, J. C.; EVANGELISTA, A. R.; SANTOS, I. P. A. Allelopathy and its interactions in the formation and management of pastures. **Boletim agropecuário**. Lavras. v. 1, n. 54, p. 1-55, 2003.

RIZZARDI, M. A.; FLECK, N. G.; VIDAL, R. A.; JUNIOR, A. M.; AGOSTINETTO. Competition for soil resources between weeds and crops. **Revista Ciência Rural**, Santa Maria, v. 31, n. 4, p. 15-22, 2001.

ROQUE, N.; BAUTISTA, H. **Asteraceae: characterization and floral morphology**. 2008.

SABELLI, R.; MOLENA, L.; PAREDE, J.; DELEO, J. P. PERSPECTIVAS 2020: Tomato. **HfBrasil**, 2020. Available at: https://www.hfbrasil.org.br/br/perspectivas-2020-tomate.aspx. Accessed on: 14 mar. 2020.

SAMPIETRO, D. A. **ALELOPATÍA: Concepto, características, metodología de estudio e importância.** Available at: https://scholar.google.com.br/scholar?hl=pt-PT&as_sdt=0%2C5&q=Alelopat%C3%ADa%3A+concepto%2C+caracter%C3%ADsticas %2C+metodolog%C3%ADa+de+estudio+e+importancia.&btnG. Accessed on: 8 feb. 2020.

SANTOS, S.; MORAES, M. L. L.; REZENDE, M. O. O.; SOUZA FILHO, A. P. S. Allelopathic potential and identification of secondary compounds in calopogonium (*Calopogonium mucunoides*) extracts using capillary electrophoresis. **Eclética Química**, São Paulo, v. 36, n. 2, p. 51-68, 2011.

SANTOS, L. R. O.; ALMEIDA, M. P.; GONÇALVES, N. R.; KROTH, B. E.; GONÇALVES, R. C. Effect of tiririca extract on corn seeds. **In:** Congresso Nacional de Iniciação Cientifica, 17º. Rondonópolis. p. 1-7, 2017.

SANTOS, D. Y. A. C. **Applied botany: secondary metabolites in plant-environment interaction**. 2015. 124 f. Thesis (Doctorate in Botany). Universidade de São Paulo - USP, São Paulo. 2015.

SANTOS, D. Q. **Potencial herbicida e caracterização química do extrato metanólico da raiz e caule do Cenchrus echinatus (Timbete)**. 2007. 70 f. Dissertation (Master in Chemistry) Universidade Federal de Uberlândia - UFU, Uberlândia, 2007.

SCHEREN, M. A.; RIBEIRO, V. M.; NOBREGA, L. H. ALELOPATHIC EFFECT OF THYRIRUM (*Cyperus rotundus* L.) ON THE DEVELOPMENT OF MAIZE PLANTULAS (*Zea mays* L.). **Varia Scientia Agrárias**, Cascavel, v. 4, n. 1, p. 105-116, 2014.

SIDRA. Levantamento sistematico da produção agrícola. 2020. Available at: https://sidra.ibge.gov.br/tabela/6588#resultado. Accessed on: 14 mar. 2020.

SIRI HARTATI, W. et al. Identification of antimicrobial compound in volatile oil of leaves of *Tagetes erecta* L. (Compositae). **Majalah Farmaci Indonesia**, v. 10, n. 1, p. 40-47, 1999.

SILVA, L.; MUELLER, S. Avaliação de coberturas vegetais no solo sobre a incidência de plantas danes e na produtividade de tomate. **Ágora: Journal of Scientific Dissemination**, Caçador, v. 17, n. 1, p. 12-19, 2010.

SILVA, D. C. **Allelopathic activity of different plant parts of *Achillea millefolium* L. and *Cymbopogon citratus* (DC) Stapf on seed germination and initial seedling development of *Lactuca sativa* L. and *Cucumis sativus* L**. 2017. 103 f. Dissertation (Master in Agronomy). Universidade Federal de Pelotas - UPF, Pelotas. 2017.
SILVA, C. P.; CAMARGO, B. F.; ROCHA, A. P. **Allelopathic effects of different species of plants from the cerrado sul-mato-grossense**. Available at: http://www.aems.edu.br/conexao/edicaoanterior/Sumario/2012/downloads/2012/saude/EF EITOS%20ALELOP%C3%81TICOS%20DE%20DIFERENTES%20ESP%C3%89CIES%2 0DE%20PLANTAS%20DO%20CERRADO%20SUL-MATOGROSSENSE.pdf. Accessed on: 9 feb. 2020.

SILVEIRA, H. R. O.; FERRAZ, E. O.; MATOS, C. C.; ALVARENGA, I. C. A.; GUILHERME, D. O.; TUFFI SANTOS, L. D.; MARTINS, E. R. Allelopathy and homeopathy in the management of tiririca (*Cyperus rotundus*). **Planta Daninha**, Viçosa, v. 28, n. 3, p. 499-506, 2010.

SOUZA FILHO, AP da S.; GUILHON, G. M. S. P.; SANTOS, L. S. Metodologias empregadas em estudos de avaliação da atividade alelopática em condições de laboratório: revisão crítica. **Planta Daninha**, Viçosa, v. 28, n. 3, p. 689-697, 2010.

THIESEN, L. A.; SCHMIDT, D.; HOLZ, E.; ALTISSIMO, B. S.; PINHEIRO, M. V. M.; HOLZ, E. Viability of *Cyperus rotundus aqueous* extract as rooting inducer in grapevine cuttings in comparison with synthetic hormones. **Acta Biológica Catarinense**, Federico Westphalen, v. 6, n. 3, p. 14-22, 2019.

VASUDEVAN, P. et al. *Tagetes*: a multipurpose plant. **Bioresource Technol**, v. 62, p. 29-35, 1997.

VIECELLI, C. A.; CRUZ-SILVA, C. T. A. Effect of seasonal variation on the allelopathic potential of Salvia. **Semina: Agrarian Sciences**, Londrina, v. 30, n. 1, p. 39-45, 2009.

VILLA, F.; FRANÇA, D. L. B.; RECH, A. L.; MOURA, C. A.; FUCHS, F. Germination of yellow passion fruit seeds in aqueous extract of tiririca and gibberellic acid. **Revista de Ciências Agroveterinárias**, Lages, v. 15, n. 1, p. 3-7, 2016.

XU, W. et al. Phytochemicals and their biological activities of plants in *Tagetes* L. **Chinese Herbal Medicines, v.** 4, n. 2, p. 103-117, 2012.

ZANATTA, J. F.; FIGUEREDO, L.; FONTANA, L. C.; PROCÓPIO, S. O. Interference of weeds in vegetable crops. **Revista da FZVA**, Uruguaiana, v. 13, n. 2, 2006.

Chapter 2

Allelopathic effect on tomato plants

Daiane Corrêa, Boris Lemes da Silva, Suelen Cristina Uber and Fabiane Nunes Silveira

Introduction

Allelopathy is defined as the capacity of plants to produce chemical substances, which when released into the environment, can influence the development of other species, both positively and negatively, through the interaction between different species (ALMEIDA apud MOLISCH, 1937).

Plants have the ability to produce primary and secondary metabolites, which are responsible for the vital functions of structure, production of new tissues, photosynthesis, synthesis of carbohydrates, proteins, lipids and nucleic acids (SANTOS, 2015), enabling their survival in different ecosystems, overcoming different factors through adaptations developed during the evolution of species (QUADROS et al., 2009).

According to Sampietro (2001), the secondary metabolite has the ability to store and produce various substances, which when released into the environment can cause beneficial effects or interfere with the development of other plants and animals. The concentration of these substances is different for each species, varying in quantity, property and even changed during its development cycle (SILVA; CAMARGO; ROCHA, 2012). The release of these substances into the environment, occurs through root exudation, volatilization, leaching and through plant decomposition (BORELLA; PASTORINI, 2009).

Among the several weed species with active principles related to allelopathy, it stands out tiririca (*Cyperus rotundus* L.), which has great importance among the cultivation of vegetables in Brazil, especially when the transition to organic production system is made (KHATOUNIAN et al., 2018). Thyririca is a plant of difficult management and control present amid the areas of crops, for having a very efficient form of vegetative propagation, being through basal bulbs, tubers and rhizomes (MIRÓ; FERREIRA; AQUILA, 1998).

In perennial crops, such as sugarcane, tiririca competes for the environment during the entire cycle, especially in the initial phases, where it releases allelopathic substances into the soil through the exudation of its roots, and that acts by inhibiting sprouting and harming the tillering of the crop, thus reducing the number of individuals per area and causing losses to the final production (PASTRE, 2006).

On the other hand, many are the studies of the applicability of extracts of tiririca as rooting inducer and growth promoter in several species, such as in grapevine (THIESEN et al., 2019). In view of this, this work aimed to evaluate the effect of extracts of tiririca (*Cyperus rotundus* L.) on the germination of seeds horticultural plants, with lettuce (*Lactuca sativa* L.) and tomato (*Lycopersicon esculentum* Mill.).

Material and Methods

The experiment was conducted at the Seed Technology Laboratory, Mato Grosso State University (UNEMAT), *campus* II Alta Floresta, during the month of February 2020.

The experimental design was entirely randomized (DIC), with four extraction forms and five different concentrations (0, 25, 50, 75 and 100%) of extracts, with four repetitions of 25 seeds each, totaling 100 seeds per treatment. To prepare the extracts, root and aerial parts of tiririca plants were collected from UNEMAT, *campus* II, in Alta Floresta - MT.

The treatments were performed using different extraction methods to obtain the extracts, being the macerated extract method and the static extract method, with different parts of the plant, being treatments obtained from extracts of the root part and extracts of the aerial part, in different concentrations of 0, 25, 50, 75, and 100%.

To obtain the macerated extract, 200g of root parts, containing roots and rhizomes, were weighed. They were then chopped with scissors and macerated by hand for 1 minute, and placed in a container with 1 liter of distilled water. The extract was left to stand for 2 hours, then filtered and used according to the different concentrations.

To obtain the macerated extract from the aerial part, 200g of the collected material was weighed, the leaves were chopped with scissors and placed in a recipient containing 1 liter of distilled water.

The static extraction method consisted in adding 1 liter of distilled water over the plant material (200 grams of aerial and root parts). This solution was left to stand for 2 hours and then the extract was filtered. As a control treatment, only distilled water was used.

The lettuce and tomato seeds used in the germination test were purchased commercially in the city of Alta Floresta - MT, being the lettuce variety "Rafaela-Americana" and the tomato variety "Gaúcho", both from Feltrin.

The germination test was performed in 9-cm diameter Petri dishes, previously autoclaved at 121 atm for 15 minutes. The substrate used was germ paper, with two sheets previously autoclaved, arranged inside the plate, which were soaked in the solution of extracts obtained or distilled water (witness) at 2.5 times the weight of the paper (g)

(BRASIL, 2009). The tests were conditioned in a BOD-type germination chamber, with a controlled temperature of 24°C, with 2°C variation and a 12-hour light photoperiod.

The evaluations were performed daily, at the same time, for seven days, and the germinated seeds were counted, considered those with emission of 2mm of primary root (HADAS, 1976). At the end of the experiment, the average length of the aerial and root parts of 5 seedlings randomly chosen from each repetition was evaluated using a ruler graduated in centimeters.

The variables analyzed were germination percentage, germination speed index, aerial part length, average root length and total length of seedlings. The variables were submitted to the analysis of variance, using the SISVAR statistical program, comparing the means of the experiment data by the Tukey test, at 5% probability.

Results and Discussion

Through the results, it was possible to verify that the static extract and macerated extract of the aerial and root parts of the thyririca (*C. rotundus L.*) proved significant $P<0.05$, on the following variables of lettuce (Lactuca *sativa L.)* and tomato (Lycopersicon *esculentum M.)* seeds, for the germination speed index, aerial part length, root length and total length of seedlings (Table 1).

Table 1- Germination percentage, germination speed index, aboveground length, root length, and total seedling length of lettuce (*Lactuca sativa* L.) seeds as a function of extraction and concentration levels of extract from the root part of *Cyperus rotundus L.*

Extraction Method	Extract Concentration				
	0	25	50	75	100
Germination Percentage (%)					
Macerated	100 aA	100 aA	99 aA	98 aA	95 aB
Static	100 aA	100 aA	99 aA	98 aA	95 aB
C.V.	3,1	3,0	3,1	3,8	3,2
Germination speed index (days)					
Macerated	22.7 bA	17.7 bA	16.5 bA	16.2 bA	12.2 bA
Static	24.2 aA	23.7 aA	24.5 aA	24.2 aA	22.2 aA
C.V.	4,3	3,5	4,1	3,7	3,9
Length of the aerial part (cm)					
Macerated	2.7 aA	1.9 aB	1.8 aB	1.8 aB	1.2 aC
Static	1.7 bA	1.5 aA	1.6 aA	1.4 ea	0.9 aB
C.V.	4,0	3,2	4,3	4,1	3,5
Root length (cm)					
Macerated	2.5 ea	2.0 aB	1.8 aB	1.6 aB	1.5 aB

Static	2.1 ea	1.7 aA	1.8 aA	1.2 aB	0.7 aC
C.V.	4,2	4,0	3,8	3,9	4,2
Total seedling length (cm)					
Macerated	5.2 aA	3.9 aB	3.6 aB	3.5 aB	2.7 aC
Static	3.9 bA	3.3 aA	3.5 aA	2.7 aB	1.7 bC
C.V.	4,8	3,5	4,1	4,5	3,7

Means followed by the same lower case letter in the column and the same upper case letter in the row do not differ, by Tukey's test, at 5% significance level.

For the evaluation of the effect of static and macerated extracts of the root part of tiririca (*C. rotundus L.*) on lettuce (*L. sativa* L.) seeds, it was observed that there was no significant influence on the percentage of seed germination. For each extract tested, the highest concentrations, with 100% of extract showed a reduction in the percentage of seed germination. These results are similar to those described by Villa et al. (2016), who observed that extracts of bulbs and leaves of tiririca at concentrations of 20, 40, 60 and 80 % caused no effect on the percentage of germination of seeds of yellow passion fruit (*Passiflora edulis* f. *flavicarpa*). The different extracts interfered in all the concentrations of the speed index of lettuce seeds germination, in which the static extract presented longer period for the germination process to occur, evidencing possible allelopathic effects. In the literature, it is possible to verify different results, such as that of Deomedesse et al. (2019), which evaluated the influence of macerated extract of tubers of tiririca (*C. rotundus* L.) on seeds of sweet corn (*Zea mays* var.*saccharata*), lettuce (*Lactuca sativa*), cucumber (*Cucumis sativus*), and cordgrass (*Ipomoea grandifolia*) seeds, observing that the extract of tiririca allowed lettuce seeds to germinate faster than the other species analyzed. Cavalcante et al. (2018), who evaluated the influence of extracts of thyrica bulbs on radish seeds, noted that extract concentrations of 15 and 25% provided a higher germination speed index. In studies such as Scheren; Ribeiro; Nobrega (2014), when evaluating the influence of extracts obtained from aerial part, bulbs and rhizome of cocoyam on the germination and development of corn seedlings, they observed that the dosages of 15 and 30% significantly reduced the rate of speed of germination of corn. Ortis et al. (2008), when verifying the effect of tiririca extracts at concentrations of 0, 25, 50, 75 and 100%, on lettuce seeds, noted a reduction in the lettuce seed germination speed index as concentrations increased. Amaral et al. (2018), when testing crude extract of tiririca potato at 100% concentration observed that, the extract not only interfered in the germination percentage of carrot (*Daucus carota* L.) seeds, but also caused retardation in its germination index.For the variable aerial part length, it was verified that the macerated and static extracts did not differ among themselves, however, they differed within each composition and dose of extract of tiririca (*C. rotundus*

L.) in the highest concentrations (Table 1). On the other hand, the static extract in the concentration of 100% showed significance (*P<0.05), differing* from the control and the other treatments. The macerated extract caused allelopathic effect, with negative action on the growth of the aerial part of lettuce seedlings, delaying and reducing their development. Similar result was found by Deomedesse et al. (2019), who observed the negative effect of the extract of tubers of tiririca on the development of the aerial part of lettuce. In the case of Lamego and Silva (2017), they noted that extract of thyrrhiza potatoes at concentrations of 25 and 50% aided in the better development of the aerial part and roots of Cabbage (*Brassica oleracea* L.), causing allelopathy with beneficial effects on plants. For the length of the roots of the seedlings, according to the results presented, both the use of macerated and static extract of root parts of tiririca did not cause significant allelopathic effects on the roots of lettuce, when compared only to the extraction method, however, within each composition of extracts, the higher doses of macerated extract considerably reduced the root length of plants when compared to the control. Scheren; Ribeiro e Nobrega (2014), verified adverse effects, in which the use of extract of bulbs and rhizomes of tiririca (*C. rotundus L.),* caused a beneficial effect on corn seedlings (*Zea mays* L.), where both aerial part and roots had a more expressive development when compared to the control.When analyzed the total length of lettuce (*L. sativa* L.) seedlings, it was observed that the concentrations (25, 50, 75 and 100%) of macerated extract of the root part of tiririca presented significant differences of the control when compared to the tested doses. In the case of the static extract, it was observed a significant difference in the concentrations of 75 and 100%, which acted negatively on the initial development of lettuce seedlings, with mean seedling size between 1.7 cm and 2.7 cm, compared to the control, whose mean was 3.9 cm.For the macerated and static extracts of the aerial parts of tiririca (*C. rotundus L.)* on lettuce seeds (*L. sativa* L.), when analyzing the variable percentage of lettuce seeds germination, the results showed that the concentrations (25, 50, 75 and 100%) of the extracts it was not possible to observe significant results (Table 2). For the dose effect, there were differences among the tested concentrations, in which the highest concentration, with 100% of extract, there were lower percentages of germination.

Table 2- Germination percentage, germination speed index, aerial part length, root length, and total seedling length of lettuce (*Lactuca sativa* L.) seeds as a function of extraction form and concentration levels of extract from the aerial part of cocoyam (*Cyperus rotundus* L.)

Extraction Method	Extract Concentration				
	0	25	50	75	100
	Germination Percentage (%)				
Macerated	100 aA	100 aA	99 aA	99 aA	91 aB
Static	100 aA	100 aA	100 aA	100 aA	95 aB

C.V.	2,9	3,1	3,0	3,0	3,6
Germination speed index (days)					
Macerated	24.5 aA	24.2 aA	24.2 aA	24.3 aA	23.7 aA
Static	24.7 aA	24.5 aA	24.5 aA	23.7 aA	23.5 aA
C.V.	3,1	3,0	3,0	3,4	3,1
Length of the aerial part (cm)					
Macerated	2.6 ea	2.5 ea	2.4 ea	2.0 aB	1.8 aB
Static	1.8 aA	1.8 aA	1.6 aA	1.5 aA	1.3 aB
C.V.	4,2	3,5	3,8	3,2	3,0
Root length (cm)					
Macerated	3.1 ea	2.7 aA	2.5 aAB	2.3 aB	2.2 aB
Static	3.1 ea	2.5 aB	2.1 aBC	2.0 aBC	1.6 aC
C.V.	2,6	3,0	3,5	3,8	4,2
Total seedling length (cm)					
Macerated	5.7 aA	5.2 aA	4.9 aA	4.3 aB	4.0 aB
Static	5.0 aA	4.3 bA	3.8 bA	3.6 aB	3.0 bC
C.V.	3,9	4,1	3,6	3,4	3,1

Means followed by the same lower case letter in the column and the same upper case letter in the row do not differ, by Tukey's test, at 5% significance level.

These results are corroborated by the experiment conducted by Andrade; Bittencourt and Vestena (2009), who observed the allelopathic effect of an extract of the fresh aerial parts of cocoyam (*C rotundus* L.) on cultivated species, noting that, at an extract concentration of 70%, there was a reduction in the percentage of germination of turnip, broccoli, cauliflower and radish seeds, but in lettuce and tomato seeds there was no reduction in the percentage of germination.

According to Gusman; Yamagush, and Vestena (2011), when testing extracts from leaves of cocoyam, black prickly pear, and milkweed on the germination of lettuce seeds, they observed that from a concentration of 10% of black prickly pear extract and from a concentration of 50% of milkweed extract the germination of lettuce seeds was reduced; however, they did not identify an allelopathic effect of the used concentrations of extracts of the aerial part of cocoyam on the germination of lettuce.

In the case of Santos et al. (2017) the action of the extract of tiririca on corn seeds was observed, and they noted that at a concentration of 50% of extract, it caused a reduction in germination and seed vigor, in addition to increasing the number of dead seeds.

For the variable aerial part length, the results presented both for the extraction methods, the macerated and static extracts of the leaves of tiririca plants, did not interfere in the development of lettuce seedlings. However, for the dose dependency effect, the 100% concentration demonstrated an allelopathic effect, resulting in seedlings with lower aerial part development. However, Filho et al. (2018), when using extracts of fresh leaves of tiririca on lettuce seeds *in vitro* observed that, at a concentration of 0.5 mg.L, it caused a better development of the aerial and root parts of the seedlings.

According to the results presented, for the length of the roots, the macerated and static extracts of tiririca leaves did not cause significant allelopathic effects on the lettuce roots when compared to the extraction method. For the concentrations, it can be evidenced that the doses of 50, 75 and 100% of extracts reduced the development of the primary radicle of the seedlings. The macerated extract of the aerial parts of tiririca at concentrations of 25, 50, 75 and 100%, did not interfere in the total length of lettuce seedlings.

When analyzing the total length of lettuce seedlings that received the static extract treatment, it was observed that in concentration of 75 and 100% presented significance by the Tukey test ($P<0.05$). Being that, the seedlings whose treatment received was 75% of static extract of tiririca leaves, differed from the concentrations of 25, 50 and 100%, but there was no significant difference when compared to the control 0%, demonstrating the allelopathic potential between the concentrations tested, reducing the initial development of the seedling after the germination process.

The seedlings that developed in static extract at the highest concentration (100%) of tiririca leaves, differed from both the control (0%) and the seedlings in concentration of 75%. There was no significant difference between the extracts at concentrations of 25 and 50%.

The results obtained from the static extract and macerated extract of the root part of cocoyam (*C. rotundus* L.) at dosages of 0, 25, 50, 75 and 100% on tomato (*L. esculentum* M.) seeds (Table 3) showed that the extract obtained by the static method showed less development of tomato seedlings.

Table 3- Germination percentage, germination speed index, aboveground length, root length, and total seedling length of tomato (*Lycopersicon esculentum* M.) seeds as a function of extraction form and concentration levels of extract from the root part of cocoyam (*Cyperus rotundus* L.)

Extraction Method	Extract Concentration				
	0	25	50	75	100
Germination Percentage (%)					
Macerated	100 aA	100 aA	98 aA	96 aA	92 aB
Static	100 aA	100 aA	100 aA	96 aA	89 aB
C.V.	3,5	3,8	3,2	30	3,0
Germination speed index (days)					
Macerated	10.2 aA	10.0 aA	9.9 aA	9.8 aA	9.6 aA
Static	8.2 bA	6.0 bA	7.0 bA	6.5 bA	5.7 bA
C.V.	3,3	3,9	3,4	4,8	3,5
Length of the aerial part (cm)					
Macerated	3.6 aA	2.7 aB	2.5 aB	2.5 aB	2.4 aB
Static	2.0 bA	1.9 bA	1.8 bA	1.3 bB	1.1 bB
C.V.	3,6	3,2	3,3	3,9	3,6
Root length (cm)					

Macerated	7.0 aA	6.2 aA	6.0 aA	5.7 aA	3.0 aB
Static	4.5 bA	3.4 bBC	3.7 bB	3.3 bBC	3.0 aC
C.V.	3,2	4,1	3,4	3,2	4,2
Total seedling length (cm)					
Macerated	10.6 aA	8.7 aB	6.7 aB	8.3 aB	5.4 aC
Static	6.6 bA	5.6 bB	5.3 bB	4.7 bB	4.2 bC
C.V.	3,0	3,0	3,1	3,5	3,2

Means followed by the same lower case letter in the column and the same upper case letter in the row do not differ, by Tukey's test, at 5% significance level.

In the first variable verified, with the percentage of germination, it was observed that both the static and macerated extracts of tiririca potatoes in concentrations of 25, 50, 75 and 100% did not differ from the witness 0%. In contrast to this result, the work done by Castro et al. (1993), demonstrates that extract of the root part of clover at concentrations of 50 and 100% interfered in the percentage of germination of tomato seeds.

For the evaluated variable germination speed index, there was no interaction between the evaluated concentrations, however, there were significant differences between the static and macerated extracts of the root part of the wormwood (*C. rotundus* L.), which may correlate the macerated extract with a greater capacity to produce allelopathic compounds, interfering in the development of the germination process, which took longer to occur.

For the variable aerial part length, it was observed that there was a significant difference between the concentrations of macerated extract, with a dose-dependent effect, in which the higher concentrations interfere more in the seedling development. On the other hand, when comparing the two extracts, it was observed that the concentrations of macerated extract (25, 50, 75 and 100%) differed from the control (0%) and the concentrations of 25 and 50% of static extract of the root part of tiririca. Pointing to an allelopathic action with negative interference, drastically reducing the development of the aerial parts of tomato seedlings.

However, in work conducted by Filho et al. (2018) observed that extracts of tiririca at concentrations of 5 mg.L and 10 mg.L corroborated the growth of the aerial part of tomatoes *in vitro*.

The concentrations of macerated extract of the root part of tiririca, did not affect the development of the roots of tomato seedlings, except for the highest dose of 100%. Regarding the static extract, the highest concentration (100%) differed from the control (0%) and the concentration of 50%, with an average of 3.0 cm, compared to the others, which averaged 4.5 (0%) cm and 3.7 (50%) cm. These results are in agreement with the work of Castro et al. (1993), where extracts of tiririca potato at concentrations of 50 and 100% drastically compromised the root development of tomato seedlings.

When analyzing the total length of tomato (*L. esculentum* L.) seedlings, it can be seen that the static extract of the root part of cocoyam, when compared with the macerated extract, at concentrations of 0, 25, 50, 75 and 100% interfered negatively in the development of tomato seedlings, reducing their total length, suggesting the potential effect of allelochemicals on the initial development of seedlings.

As shown in Table 4, for the static and macerated extract of the aerial part of tiririca (*C. rotundus* L.) on tomato (*L. esculentum* M.) seeds, it was possible to observe that the concentrations of 25, 50, 75 and 100% of macerated and static extract did not present significant results for the germination percentage when compared among the extraction methods. For the concentrations, the highest ones provided a reduction in the percentage of seed germination.

Table 4- Germination percentage, germination speed index, aboveground length, root length, and total seedling length of tomato (*Lycopersicon esculentum* M.) seeds as a function of extraction and concentration levels of extract from the aerial part of cocoyam (*Cyperus rotundus* L.)

Extraction Method	Extract Concentration				
	0	25	50	75	100
Germination Percentage (%)					
Macerated	100 aA	100 aA	98 aA	95 aA	91 aB
Static	100 aA	100 aA	98 aA	94 aB	90 BC
C.V.	2,9	3,0	3,0	3,1	3,2
Germination speed index (days)					
Macerated	10.2 aA	9.7 aA	9.6 aA	9.2 aA	8.7 aA
Static	10.0 aA	10.0 aA	10.0 aA	9.7 aA	6.5 bB
C.V.	3,0	3,3	3,5	3,9	3,5
Length of the aerial part (cm)					
Macerated	3.4 ea	3.4 ea	3.3 aA	3.3 aA	2.5 ea
Static	3.2 aA	2.9 aA	2.9 aA	2.7 aA	2.5 aB
C.V.	3,1	3,9	3,5	3,5	3,7
Root length (cm)					
Macerated	6.6 aA	6.6 aA	5.7 aA	4.4 bB	4.3 aB
Static	6.4 aA	6.2 aA	5.9 aA	5.6 aA	3.9 aB
C.V.	3,0	3,8	3,9	4,8	4,1
Total seedling length (cm)					
Macerated	10.1 aA	10.3 aA	9.1 aA	7.7 aB	6.8 aB
Static	9.6 aA	9.1 bA	8.8 aA	8.4 aA	6.0 aB
C.V.	3,8	3,3	3,1	3,3	3,0

Means followed by the same lower case letter in the column and the same upper case letter in the row do not differ, by Tukey's test, at 5% significance level.

Similar results were found by Andrade; Bittencourt and Vestena (2009), when testing aqueous extract of dried leaves of tiririca at concentrations of 10, 30, 50, 70, 90, 100% on seeds of cultivated species, they observed that the tested concentrations did not interfere in the percentage of germination of tomato seeds. On the other hand, Borella and Pastorini

(2009) observed that extracts of umbu (*Phytolacca dioica* L.) leaves at a concentration of 8% completely inhibited the germination of tomato and prickly pear seeds.

The macerated extract did not influence the germination speed index of tomato seeds. When analyzing the static extract, it is noted that the 100% concentration benefited the seeds, which had a faster germination than the control (0%), and the other seeds on different extract concentrations of 25, 50 and 70%. Scheren; Ribeiro and Nobrega (2014), observed the influence of extracts of root parts and aerial part of tiririca on corn seeds, noting that, extract concentrations of 15 and 30% significantly reduced the speed of germination of corn seeds.

The aerial part length of tomato seedlings, it is found that at both concentrations of macerated and static extract, did not significantly influence the development of tomato seedlings.

Andrade; Bittencourt and Vestena (2009), found in their work that extract of the aerial part of tiririca in crude concentration (100%), impaired the development of the aerial part of tomato seedlings. Gusman; Yamasgushi and Vestena (2011), observed that both extracts of aerial parts of milkweed, black prickly pear and thyrika at concentrations of 10, 30, 50 and 70% drastically reduced germination, aerial part and root development of tomato.

The root length of tomato seedlings was positively influenced by the static extract at the 75% concentration when compared to the macerated extract, allowing the seeds to get a better root development.

When analyzing the total length of tomato seedlings it appears that the macerated extract of the aerial parts of tiririca did not influence the length of the seedlings, on the other hand, when evaluating the difference between the concentrations of 25% and 100% it is observed significance between the concentrations, but there is no significant difference when compared to the control (0%) and concentrations of 50 and 75%, reducing the growth and development of the seedling.

Conclusion

The static extracts of the root part and the aerial part of the weed (*C. rotundus* L.) were very representative in concentrations of 75 and 100%, causing allelopathic effects, with negative interference on the variables germination speed index, aerial part length, root length and total length of horticultural seedlings, such as lettuce (*L. sativa* L.) and tomato (*L. esculentum* M.), the latter being shown to have a greater interference effect when compared to the macerated extraction method.

References

ALENCAR, G. Unpublished study identifies major weeds to tomatoes intended for industry. **EMBRAPA**, 2018. Available at: https://www.embrapa.br/ busca-de-noticias/-/noticia/34802765/estudo-inedito-identifica-principais-plantas-daninhas-ao-tomate-destinado-a-industria. Accessed on: 14 Mar. 2020.

ALMEIDA, F. S. Allelopathic effects of plant residues. **Revista Pesquisa Agropecuária Brasileira,** Brasília, v. 10, p. 221-236, 1991.

ALVES, J. N. Chemical Characterization of Dichloromethane Extracts of *Origanum majorana* L. in the Inhibition of *Panicum maximum*. 2009. 96 f. Master (Post Graduation Chemistry) - Universidade Federal de Uberlândia - UFU, Uberlândia. 2009.

ALVES, M. C. S.; FILHO, S. M.; INNECCO, R.; TORRE, S. B. Allelopathy of volatile extracts on seed germination and root length of lettuce. **Pesquisa Agropecuária Brasileira**, Brasília, v. 39, n. 11, p. 1083-1086, 2004.

AMARAL, M. C. A.; BANDEIRA, A. S.; PORTO, J. S.; ÁVILA, J. S.; SANTOS, R. K. A.; MORAIS, O. M. Avaliação do efeito allelopático de extrato aquoso de tiririca sobre a germinação de sementes de carrot. **Cadernos de Agroecologia**, Seropédica, v. 13, n. 1, p. 1-7, 2018.

ANDRADE, H. M.; BITTENCOURT, A. H. C.; VESTENA, S. Allelopathic potential of Cyperus rotundus L. on cultivated species. **Ciência e Agrotecnologia**, Lavras, v. 33, n. SPE, p. 1984-1990, 2009.

BARATELLI, T. G. **Study of plant allelopathic properties: investigation of allelochemical substances in *Terminalia catappa* L. (Combretaceae).** 2006. 206 f. Thesis (Doctorate in Natural Products Chemistry) - Universidade Federal do Rio de Janeiro - UFRJ, Rio de Janeiro. 2006.

BORELLA, J.; PASTORINI, L. H. Allelopathic influence of Phytolacca dioica L. on germination and initial growth of tomato and black pepper. **Revista Biotemas**, Florianópolis, v. 22, n. 3, p. 67-75, 2009.

BORELLA, J.; TUR, C. M.; PASTORINI, L. H. Allelopathy of aqueous extracts of *Duranta repens* on germination and initial growth of Lactuca sativa and Lycopersicum esculentum. **Revista Biotemas**, Florianópolis, v. 23, n. 2, p. 13-22, 2010.

BRASIL. Ministry of Agriculture, Livestock and Supply. **Regras para análises de sementes**. Brasília: MAPA/ACS, 2009. 398p.

BRZEZINSKI, C. R.; ABATI, J.; GELLER, J.; WERNER, F.; ZUCARELI, C. Production of American lettuce cultivars under two cropping systems. **Revista Ceres**, Viçosa, v. 64, n. 1, 2017.

CASTRO, P. R. C.; RODRIGUES, J. D.; MORAES, M. A.; CARVALHO, V. L. M. Allelopathic effects of some plant extracts on tomato (*Lycopersicon esculentum* Mill. cv. Santa Cruz) germination. **Planta Daninha**, Piracicaba, v. 6, n. 2, p. 79-85, 1983.

CAVALCANTE, J. A.; LOPES, K. P.; PEREIRA, N. A. E.; SILVA, J. G.; PINHEIRO, R. M.; MARQUES, R. L. L. Aqueous extract of thyrrhiza bulbs on germination and initial growth of radish seedlings. **Revista Verde de Agroecologia e Desenvolvimento Sustentável**, Pelotas, v. 13, n. 1, p. 39-44, 2018.

COMIOTTO, A. **Allelopathic potential of different plant species on the physiological quality of rice seeds and lettuce achenes and growth of rice and lettuce seedlings**. 2006. 42 f. Dissertation (Master of Science) Universidade Federal de Pelotas - UPF, Pelotas. 2006.

CONAB. **Boletim hortigranjeiro**. Brasília. v. 5, n. 4, April 2019.

CORSATO, J. M.; FORTES, A. M. T.; SANTORUM, M.; LESZCZYNSKI, R. Allelopathic effect of aqueous extract of sunflower leaves on the germination of soybean and black walnut. **Semina: Agrarian Sciences**, Londrina, v. 31, n. 2, p. 353-360, 2010.

CREMONEZ, F. E.; CREMONEZ, P. A.; CAMARGO, M. P.; FEIDEN, A. Principal plants with allelopathic potential found in Brazilian agricultural systems. **Acta Iguazu**, Cascavel, v. 2, n. 5, p. 70-88, 2013.

DEOMEDESSE, C. C.; MENESES, N. B.; SOUSA, G. O.; SILVA, T. S.; CRUZ, G. A. Allelopathic effects of tiririca extract on the germination of sweet corn, lettuce, cucumber, and string. **MAGISTRA**, Cruz das Almas, v. 30, p. 323-330, 2019.

DIAS, J. F.G. **Applied allelopathic study of Aster lanceolatus, Willd.** 2005. 46f. Dissertation (Master in Pharmaceutical Sciences). Universidade Federal do Paraná, Curitiba. 2005.

DIAS, J. F. G.; CÍRIO, G. M.; MIGUEL, M. D.; MIGUEL, O. G. Contribuição ao estudo alelopático de Maytenus ilicifolia Mart. ex Reiss., Celastraceae. **Revista Brasileira de Farmacognosia**, João Pessoa, v. 15, n. 3, p. 220-223, 2005.

DUSI, A. N. A cultura do tomatoiro (para mesa). **EMBRAPA-SPI. Coleção plantar**, 1993.

Allelopathic Studies Related to Coffee. Porto Velho: EMBRAPA, n. 54, p. 26. 2001. *ISSN 0103-9865.*

FERREIRA, A. G.; AQUILA, M. E. A. Allelopathy: an emerging area of ecophysiology. **Revista Brasileira de Fisiologia Vegetal**, Brasília v. 12, n. 1, p. 175-204, 2000.

FILHO, R. F.; SANTOS, B. V. B.; LOPES, T. B.; SOUZA, B. N.; SOUZA, B. R. Utilizacao de extrato de folhas de Tirirca in *vitro* na cultura da Alface. **In:** Congresso Nacional de Meio Ambiente, 15º. Poços de Caldas. p. 1-5, 2018.

FORMAGIO, A.; MASETTO, T. E.; BALDIVIA, D. S.; VIEIRA, M. C.; ZÁRATE, N. A. H.; PEREIRA, Z. V. Potencial alelopático de cinco espécies da família Annonaceae. **Revista Brasileira de Biociências**, Porto Alegre, v. 8, n. 4, 2010.

FONTANÉTTI, A.; CARVALHO, G. J.; GOMES, L. A. A.; DE ALMEIDA, K.; DE MORAES, S. R. G.; DUARTE, W. F. Efeito alelopático da adubação verde no controle de tiririca (Cyperus rotundus L.). **Cadernos de Agroecologia**, Seopédica, v. 2, n. 1, 2007.

FRANÇA, A. C.; SOUZA, I. F.; SANTOS, C. C.; OLIVEIRA, E. Q.; MARTINOTTO, C. Allelopathic activities of neem on the growth of sorghum, lettuce and black walnut. **Revista Ciência e Agrotecnologia**, Lavras, v. 32, n. 5, p. 1374-1379, 2008.

FUMAGALI, E.; GONÇALVES, R. A. C.; MACHADO, M. F. P. S.; VIDOTI, G. J.; OLIVEIRA, A. J. B. Production of secondary metabolites in plant cell and tissue culture: The example of the *Tabernaemontana* and *Aspidosperma* genera. **Revista Brasileira de Farmacognosia**, João Pessoa, v. 18, n. 4, p. 627-641, 2008.

GATTI, A. B. **Allelopathic activities of cerrado species**. 2008. 138 f. Thesis (PhD in Ecology and Natural Resources) Universidade Federal de São Carlos - UFSCar, SP. 2008.

GUSMAN, G. S.; YAMAGUSHI, M. Q.; VESTENA, S. Allelopathic potential of aqueous extracts of *Bidens pilosa* L., *Cyperus rotundus L.* and *Euphorbia heterophylla* L.. **Iheringia. Botany Series,** Porto Alegre, v. 66, n. 1, p. 87-98, 2011.

HADAS, A. Water uptake and germination of leguminous seeds under changing external water potencial in osmotic solution. **Journal Express Botany**, Saint Louis, v. 27, p. 480-489, 1976.

KHATOUNIAN, C. A.; OLIVEIRA, D. A. M.; FERREIRA, T. M.; DUPRE, M.; MERIANNE, H. Distribution of tubers of Tiririca (*Cyperus rotundus* L.) in the soil profile and its implications for conversion to organic agriculture of urban gardens. **Scientia Plena**, Piracicaba, v. 14, n. 9, 2018.

LAMEGO, K. B.; DA SILVA, J. Allelopathic potential of the aqueous extract of the tubers of tiririca. **South American**, Ji-Paraná, v. 4, n. 2, p. 84-93, 2017.

MARTINELLI, V. A.; SILVA, V. N. Allelopathic effect of rye on germination and growth of beet seedlings. **Agrarian Academy Magazine**, Goiânia, v. 5, n. 9; p. 1-9, 2018.

MIRÓ, Cíntia Pilotti; FERREIRA, Alfredo Gui; AQUILA, Maria Estefânia Alves. Allelopathy of yerba mate (*Ilex paraguariensis*) fruits on corn development. **Pesquisa Agropecuária Brasileira**, Brasilia, v. 33, n. 8, p. 1261-1270, 1998.

MOGHARBEL, A. D. I.; MASSON, M. L. Perigos associados ao consumo da alface (*Lactuca sativa*), in natura. **Alimentação Nutritiva**, Araraquara, v. 16, n. 1, p. 83-88, 2005.

MOLISCH, H. **The influence of one plant on another.** Published by Gustav Fischer, 1937.

MORALES-PAYAN, J. P.; STALL, W. M.; SHILLING, D. G.; CHARUDATTAN, R.; DUSKY, J. A.; BEWICK, T. A. Above-and belowground interference of purple and yellow nutsedge (*Cyperus* spp.) with tomato. **Weed science**, Amsterda, v. 51, n. 2, p. 181-185, 2003.

NAVARRO, L. F. F.; SILVA, M. S.; MOECKE, U. F. R. **Efficiency of the organic extract of "Cyperus rotundus as a rooter in the propagation of Corymbia citriodora".** 2017. 55 f. Monograph (Bachelor in Agronomy) Centro Universitário Católico Salesiano *Auxilium* - UniSALESIANO, Lins. 2017.

ORTIZ, T. A.; MORITA, D. A. S.; COSMO, D. A.; SELEGUINI, A.; ARIERA, C. D.; HORA, R. C. 2008. Germination of lettuce seeds as a function of concentrations of aqueous extract of tiririca. **Revista Horticultura Brasileira**, Umuarama, v. 26, n. 2, p. 1-5, 2008.

PASTRE, W. **Control of thyrrhiza (*cyperus rotundus* L.) with sulfentrazone and flazasulfuron applied alone and in mixture in sugarcane crop.** 2006. 66 f. Dissertation (Master in Tropical and Subtropical Agriculture) Agronomic Institute - IAC, Campinas. 2006.

PITELLI, R. A. Competição e controle das plantas weeds em áreas agrícolas. **Série técnica IPEF**, Piracicaba, v. 4, n. 12, p. 1-24, 1987.

QUADROS, B. R.; TANAKA, A. A.; CARDOSO, A. I. I.; RIGOTTI, M.; SANTOS, R. F. Allelopathy of plant extracts of pelargonium graveolens and cyperus rotundus on lettuce seed germination. **Revista Horticula Brasileira**, Botucatu, v. 27, n. 2, 2007.

QUEIROZ, A. A.; CRUVINEL, V. B.; FIGUEIREDO, K. Production of American lettuce as a function of fertilization with organomineral. **Enciclopédia Biosfera, Centro Científico Conhecer**, Goiânia, v. 14, n. 25, p. 1-12, 2017.

REGHIN, M. Y.; PURÍSSIMO, C.; DALLA PRIA, M.; FELTRIM, A. L.; FOLTRAM, M. A. Técnicas de cobertura do solo e de proteção de plantas no cultivo da alface. **Horticultura Brasileira**, Ponta Grossa, v. 20, n. 2, p. 1-10, 2002.

REZENDE, C. de P.; PINTO, J. C.; EVANGELISTA, A. R.; SANTOS, I. P. A. Allelopathy and its interactions in the formation and management of pastures. **Boletim agropecuário.** Lavras. v. 1, n. 54, p. 1-55, 2003.

RIZZARDI, M. A.; FLECK, N. G.; VIDAL, R. A.; JUNIOR, A. M.; AGOSTINETTO. Competition for soil resources between weeds and crops. **Revista Ciência Rural**, Santa Maria, v. 31, n. 4, p. 15-22, 2001.

ROQUE, N.; BAUTISTA, H. **Asteraceae: characterization and floral morphology**. 2008.

SABELLI, R.; MOLENA, L.; PAREDE, J.; DELEO, J. P. PERSPECTIVAS 2020: Tomato. **HfBrasil**, 2020. Available at: https://www.hfbrasil.org.br/br/perspectivas-2020-tomate.aspx. Accessed on: 14 mar. 2020.

SAMPIETRO, D. A. **ALELOPATÍA: Concepto, características, metodología de estudio e importância.** Available at: https://scholar.google.com.br/scholar?hl=pt-PT&as_sdt=0%2C5&q=Alelopat%C3%ADa%3A+concepto%2C+caracter%C3%ADsticas %2C+metodolog%C3%ADa+de+estudio+e+importancia.&btnG. Accessed on: 8 feb. 2020.

SANTOS, S.; MORAES, M. L. L.; REZENDE, M. O. O.; SOUZA FILHO, A. P. S. Allelopathic potential and identification of secondary compounds in calopogonium (*Calopogonium mucunoides*) extracts using capillary electrophoresis. **Eclética Química**, São Paulo, v. 36, n. 2, p. 51-68, 2011.

SANTOS, L. R. O.; ALMEIDA, M. P.; GONÇALVES, N. R.; KROTH, B. E.; GONÇALVES, R. C. Effect of tiririca extract on corn seeds. **In:** Congresso Nacional de Iniciação Cientifica, 17º. Rondonópolis. p. 1-7, 2017.

SANTOS, D. Y. A. C. **Applied botany: secondary metabolites in plant-environment interaction**. 2015. 124 f. Thesis (Doctorate in Botany). Universidade de São Paulo - USP, São Paulo. 2015.

SANTOS, D. Q. **Potencial herbicida *e* caracterização química do extrato metanólico da raiz *e caule do Cenchrus echinatus* (Timbete)**. 2007. 70 f. Dissertation (Master in Chemistry) Universidade Federal de Uberlândia - UFU, Uberlândia, 2007.

SCHEREN, M. A.; RIBEIRO, V. M.; NOBREGA, L. H. ALELOPATHIC EFFECT OF THYRIRUM (*Cyperus rotundus* L.) ON THE DEVELOPMENT OF MAIZE PLANTULAS (*Zea mays* L.). **Varia Scientia Agrárias**, Cascavel, v. 4, n. 1, p. 105-116, 2014.

SIDRA. Levantamento sistematico da produção agrícola. 2020. Available at: https://sidra.ibge.gov.br/tabela/6588#resultado. Accessed on: 14 mar. 2020.

SILVA, L.; MUELLER, S. Avaliação de coberturas vegetais no solo sobre a incidência de plantas danes e na produtividade de tomate. **Ágora: Journal of Scientific Dissemination**, Caçador, v. 17, n. 1, p. 12-19, 2010.

SILVA, D. C. **Allelopathic activity of different plant parts of *Achillea millefolium* L. and *Cymbopogon citratus* (DC) Stapf on seed germination and initial seedling** development of *Lactuca sativa* **L.** and *Cucumis sativus* **L.** 2017. 103 f. Dissertation (Master in Agronomy). Universidade Federal de Pelotas - UPF, Pelotas. 2017.

SILVA, C. P.; CAMARGO, B. F.; ROCHA, A. P. **Allelopathic effects of different species of plants from the cerrado sul-mato-grossense**. Available at: http://www.aems.edu.br/conexao/edicaoanterior/Sumario/2012/downloads/2012/saude/EF EITOS%20ALELOP%C3%81TICOS%20DE%20DIFERENTES%20ESP%C3%89CIES%2 0DE%20PLANTAS%20DO%20CERRADO%20SUL-MATOGROSSENSE.pdf. Accessed on: 9 feb. 2020.

SILVEIRA, H. R. O.; FERRAZ, E. O.; MATOS, C. C.; ALVARENGA, I. C. A.; GUILHERME, D. O.; TUFFI SANTOS, L. D.; MARTINS, E. R. Allelopathy and homeopathy in the management of tiririca (*Cyperus rotundus*). **Planta Daninha**, Viçosa, v. 28, n. 3, p. 499-506, 2010.

SOUZA FILHO, AP da S.; GUILHON, G. M. S. P.; SANTOS, L. S. Metodologias empregadas em estudos de avaliação da atividade alelopática em condições de laboratório: revisão crítica. **Planta Daninha**, Viçosa, v. 28, n. 3, p. 689-697, 2010.

THIESEN, L. A.; SCHMIDT, D.; HOLZ, E.; ALTISSIMO, B. S.; PINHEIRO, M. V. M.; HOLZ, E. Viability of *Cyperus rotundus aqueous* extract as rooting inducer in grapevine cuttings in comparison with synthetic hormones. **Acta Biológica Catarinense**, Federico Westphalen, v. 6, n. 3, p. 14-22, 2019.

VIECELLI, C. A.; CRUZ-SILVA, C. T. A. Effect of seasonal variation on the allelopathic potential of Salvia. **Semina: Agrarian Sciences**, Londrina, v. 30, n. 1, p. 39-45, 2009.

VILLA, F.; FRANÇA, D. L. B.; RECH, A. L.; MOURA, C. A.; FUCHS, F. Germination of yellow passion fruit seeds in aqueous extract of tiririca and gibberellic acid. **Revista de Ciências Agroveterinárias**, Lages, v. 15, n. 1, p. 3-7, 2016.

ZANATTA, J. F.; FIGUEREDO, L.; FONTANA, L. C.; PROCÓPIO, S. O. Interference of weeds in vegetable crops. **Revista da FZVA**, Uruguaiana, v. 13, n. 2, 2006.

Chapter 3

Chapter 3

Allelopathic effect on vegetable plants

Daiane Corrêa, Michele Fernanda Bortolini, Suelen Cristina Uber and Fabiane Nunes Silveira

Introduction

Allelopathy is defined as the chemical interference that one plant exerts on another, which can be positive or negative. The term *allelopathy* derives from the Greek *allelon*, mutual, and *pathos*, damage, due to the production of chemical compounds released into the environment, originating from secondary metabolism, which are called allelochemicals (RICE, 1984).

Allelopathy is widely known as an important ecological mechanism that plays a role in crop plant communities, natural vegetation, commercial crops, and invasive plants in different agroecosystems (BARATELLI, 2006).

It can also be defined as processes that produce secondary metabolites, acting as phytotoxic compounds, inhibiting or promoting some biochemical or physiological processes, according to the variation in the concentration of one or more allelochemicals in other plants or organisms (FERREIRA; AQUILA, 2000).

The chemical compounds that have allelopathic activity are present in all plants and distributed in their tissues, such as in the leaves, flowers, fruits, seeds, stems and roots and the allelopathic phenomena are not caused by a single chemical substance. Generally the interference occurs by the combined action of a number of allelochemicals, and this impact is also influenced by other factors, such as environmental stress (ALVES, 2003).

One of the most explored aspects of allelopathy is its role in organic agriculture, because several species, besides the cultivated ones present in agroecosystems, can exert an allelopathic influence on crops, consequently interfering in their germination and development.

In some cultivated species it has been observed that its residues can control or even suppress the emergence and development of some species, both in cultivated and invasive plants (HUANG; CHOU, 2005).

The most frequent studies on allelopathy are related to the effects of plant extracts, as natural herbicides, on germination and growth of other plants (PEREZ; LIMA, 2004). In general, one can observe these effects on germination and root length, but germination is less sensitive to allelochemicals than radicle growth (TAIZ; ZEIGER, 2009).

Among the plants used in agroecological systems, there is the ornamental carnation plant (*Tagetes* sp.), belonging to the Asteraceae family, originally from Mexico, with a strong aroma and an eye-catching color (FERRAZ; FREITAS, 1995).

Clove is used with nematicidal and insecticidal functions in organic farming, but there are no reports of studies in Brazil on its allelopathic potential as a way to control weeds (CHAMORRO et al., 2008).

Therefore, even though there is a considerable wealth of literature on allelopathy and methodologies for extraction forms in biological assays, the literature is incipient for the extraction of secondary compounds from clove. Thus, research is needed to analyze the interference caused by allelopathy in agroecological systems of this plant. Thus, it is necessary to identify the effects and allelopathic potential in plants.

Material and Methods

The experiments were conducted in the Biotechnology Laboratory of the Pontifical Catholic University of Paraná (PUCPR), Toledo *campus,* from February to March 2010. *Tagetes patula* L. (carnation) were identified and registered in the Herbarium of UNIOESTE - UNOP, Cascavel campus, PR, Brazil, under number 5982.

Fresh carnation petals were used for the experiment, grown in 2009/2010 in soil classified as typical red dystrophic Latosol of clayey texture, which was fallow, without the presence of chemical residues, in the city of Céu Azul, PR, whose coordinates are 53º 44' 05" W and 25º 01'36" S, at 550m altitude. The rubin seeds were also collected in this location.

The fodder turnip seeds, cultivar IPR 116, were purchased from Instituto Agronômico do Paraná - IAPAR, the lettuce seeds variety "Grads Rapids" (bioindicator), watermelon variety "Crinsom Sweet" and melon variety "Cantaloupe", both from the brand TopSeed, were purchased commercially in the city of Toledo-Pr.

The experiment consisted of an assay testing different forms of extraction, through the grinding method and static and with different concentrations (0, 20, 40, 60, 80 and 100%) of extract. To obtain the aqueous extracts from the petals of the plant, the extraction method was tested by grinding 200g of clove petals with the help of a blender and adding 1 liter of distilled water, and then filtered with filter paper (Whatmann 02).

The static extraction method consisted in adding 1 liter of distilled water over the plant material (200 grams of petals). This solution remained at rest for 24 hours, with no incidence of light, and was then filtered. The pH of the aqueous extracts was analyzed using a phagameter, in which the triturated extract had a pH of 5.7 and the static extract had a pH of 5.3. As a control treatment, only distilled water was used.

The germination test was performed in Petri dishes, 9 cm in diameter, previously autoclaved at 121 atm for 15 minutes. The substrate germ paper, arranged inside the plate, were soaked by the solution of extracts obtained or distilled water (witness) at 2.5 times the weight of the paper (g) (BRASIL, 2009).

For lettuce, rutabaga, and turnip seeds, Petri dishes with 25 seeds each were used. For the germination test on melon and watermelon seeds, germ paper rolls were used, also moistened with the different solutions of the extract in the same proportion, containing 50 seeds each.

The rube seeds were previously disinfected from a 1% sodium hypochlorite solution for 5 minutes and the turnip, melon, watermelon and lettuce seeds were previously treated with fungicide (Tebuconazole 0.2 g.i.a.).

The tests were installed in a BOD-type germination chamber, with a controlled temperature of 25°C, with 2°C variation and a 12-hour light photoperiod. Evaluations were performed daily, at the same time, for 16 days, counting the germinated seeds, considered those with emission of 2mm of primary root (HADAS, 1976), at the end of the experiment was also evaluated the average length of 5 roots chosen at random in each repetition, using a ruler graduated in centimeters.

The variables analyzed were germination percentage, mean germination time, mean germination speed and mean root length. The experimental design was entirely randomized (DIC), in a factorial scheme (2 x 6), with two forms of extraction and six different concentrations, with four repetitions of 25 or 50 seeds each, according to the seeds tested. The variables were submitted to variance analysis, using the statistical program SISVAR, comparing the means of the experiment data by Tukey's test, at 5% probability.

Results and Discussion

According to the results obtained for the lettuce seeds, it can be observed that there was interaction between the extraction form and the extract concentration. The lettuce seeds germination percentage suffered interference from the extraction form, where the triturated and static extraction differed significantly from the concentration of 40%, as shown in Table 1.

For the extracts obtained from the grinding method, there were significant differences in the highest concentrations (80 and 100%), which did not differ among themselves. The extracts obtained through the static form differed statistically from each other from the concentration of 60%, obtaining averages of 87, 27 and 12% of germination respectively for 60, 80 and 100% of extract concentration.

Table 1- Germination percentage, mean germination speed, mean germination time, and mean root length of lettuce (*Lactuca sativa* L.) seeds as a function of extraction form and concentration levels of aqueous extract of clove (*Tagetes patula* L.)

	Germination Percentage (%)						
Extraction	Extract Concentration						
Method	0	20	40	60	80	100	Averages
Crushed	100 aB	100 aB	100 bB	100 bB	94 bA	93 Ba	97,8
Static	100 aD	100 aD	96 aD	87BC	27 aB	12 Aa	70,3
AVERAGE	100	100	98	93,5	60,5	52,5	
	Average Germination Speed (days)						
Crushed	0.070aB	0.070 aB	0.070 aB	0.070bB	0.060 yA	0.070 Bb	0,068
Static	0.070 aB	0.070 aB	0.070 aB	0.067 aB	0.060 yA	0.060 yA	0,066
AVERAGE	0,070	0,070	0,070	0,068	0,060	0,065	
	Average Germination Time (days)						
Crushed	13.17 aA	13.81aB	13.91aB	13.86 bB	14.68 bC	14.08 bB	13,91
Static	13.17 aA	13.45 bA	13.36 bA	14.25 aB	15.00 aC	15.12 BC	14,05
AVERAGE	13,17	13,63	13,63	14,04	14,84	14,60	
	Average Root Length (cm)						
Crushed	3.70 aC	2.95 bA	3.13 bAB	3.18 bAB	3.45 bBC	3.35bABC	3,29
Static	3.51 aC	2.63 aC	2.48 aC	1.38 aB	1.05aAB	0.88 AA	1,88
AVERAGE	3,29	2,79	2,80	2,28	2,25	2,11	

Means followed by the same lower case letter in the column and the same upper case letter in the row do not differ, by Tukey's test, at 5% significance level.

With the present results, it can be observed that in the static extraction there was a significant reduction in the average germination percentage, reaching 12%; however, these results were obtained in experiments carried out under controlled conditions, because in natural conditions in the field, it is not possible to obtain such high concentrations of extract in the environment.

Similar results to the present study were obtained by Baratto et al. (2008), in which crushed extracts of guaco (*Mikania laevigata*) leaves (Asteraceae), the same family as clove, inhibited in 100% the germination of lettuce seeds.

For the different concentrations tested, it can be observed that as the extract concentration increased, there was a reduction in the germination percentage, ranging from 12 to 100% germination.

According to reports by Ilori et al. (2010), in tests performed with extracts obtained from static extractions of *Chromolaena odorata* (arnica-do-mato), *Helianthus annus* (sunflower) and *Tithonia diversifolia* (sunflower-mexican), plants of the Asteraceae family.

They also showed negative interference on germination and radicle growth of *Vigna unguiculata* (cowpea), since many species of the Asteraceae family have secondary metabolites in their composition that are used as sources of bioactive molecules.

For the average speed of germination, it was also verified and interaction between the tested factors, and there was difference between the forms of extraction only in the concentration of 60% and 100%. The lowest average speed of germination obtained in the static extract was 0.060 day in the concentrations of 80 and 100%, differing from the others.

For the average time variable, which presented interaction among the tested factors, it can be observed that there was interference of the extraction form and the extract concentrations. The control treatment (13.17 days) and the 20% concentration of crushed extract (13.81 days) differed from each other and from the 80% extract concentration.

For the static extraction form, it can be observed that the highest average time was recorded for the concentration of 80 and 100%, with 15.00 and 15.12 days. Through these results it can be observed that the static extract obtained the lowest indices of average time of germination in the highest concentrations (60, 80, and 100%), with 14.25; 15.00, and 15.12 days in relation to the extraction in triturated form.

For the average root length, which also registered interference among the tested factors, it was verified that the crushed extract differed statistically from the static extract in all concentrations. For the static extraction, the control treatment and the concentrations of 20 and 40% were equal to each other and differed from the higher concentrations (60, 80, and 100%). In this extraction it can be observed that as the concentration increased, a reduction in average root length occurred.

These results are in agreement with the tests performed by Kil and Yun (1992), used aqueous extracts of *Artemisia princeps* (Asteraceae), obtained statically, which reduced the germination percentage and root length in seeds of *Lactuca sativa* L., *Achyranthes japonica*, *Artemisia princeps*, *Oenothera odorata*, *Plantago asiatica* L., *Zoysia japonica*, *Echinochloa crus-galli* L., *Hordeum vulgare* and *Chrysanthemum boreale* at different concentrations.

According to Kil et al. (2000), plants of the genus *Tagetes* have been studied in relation to the production of secondary compounds, due to its great importance in agroecological systems, because the macerated extracts of *Tagetes minuta* negatively interfered in the development, in the average root length and in the formation of root hairs in bioassays using lettuce (*Lactuca sativa* L.) and lotus (*Lotus corniculatus)* seeds.

According to Kil et al., (2002), the aqueous extracts of *T. minuta* inhibited germination and altered root growth patterns of lettuce (*Lactuca sativa L.)* and broom (*Lotus corniculata* L.) seeds, but the greatest phytotoxic effect was observed in the radicles, both by reducing the average length and by necrosis at their extremities.

Through the results obtained for the turnip rape seeds, for the germination percentage, it can be observed that there was no interaction among the evaluated factors. For the extract concentration factor, 60, 80 and 100% differed from the control treatment, demonstrated in the results presented as shown in Table 2.

For the extraction method there was a significant difference between static and triturated extraction, with static extraction having the lowest average, with 68.8% germination, while triturated extraction had an average of 78.1%.

As described by Silva et al. (2009), these results are confirmed by the study performed with sunflower extracts (*Helianthus annus* L.), of the same family as *Tagetes* sp. which showed allelopathic potential, inhibiting the germination of seeds of weeds, *Bidens pilosa* and *Amaranthus hybridus,* as well as of *Lactuca sativa* L.

Table 2- Germination percentage, mean germination speed, mean germination time and mean root length of rape (*Raphanus sativus* L.) seeds as a function of extraction form and concentration levels of aqueous extract of clove (*Tagetes patula* L.)

Germination Percentage (%)							
Extraction	Extract Concentration						
Method	0	20	40	60	80	100	Averages
Crushed	90	84	80	77	73	65	78,1b
Static	84	71	74	63	61	60	68,8a
AVERAGE	87 C	77,5 B	77 B	70 AB	67 A	62,5 A	
Average Germination Speed (days)							
Crushed	0,071	0,073	0,070	0,067	0,069	0,070	0,070 a
Static	0,069	0,071	0,071	0,068	0,069	0,068	0,070 a
AVERAGE	0.070 AB	0,072 B	0.070 AB	0,068 A	0.070 AB	0.069 AB	
Average Germination Time (days)							
Crushed	13,95	13,72	14,17	14,79	14,18	14,15	14,16 a
Static	14,32	14,32	14,08	14,54	14,34	14,59	14,30 a
AVERAGE	14.14 AB	13,81 A	14.12 AB	14,67 B	14.26 AB	14.37 AB	
Average Root Length (cm)							
Crushed	5.40 aC	3.40 aB	2.85 aAB	2.92 aAB	2.62 aA	2.47 aA	3,27
Static	5.40 aC	3.37 aAB	3.57 bB	3.47 bAB	2.90 aAB	2.85 yA	3,42
AVERAGE	4,9	3,38	3,21	3,19	2,76	2,66	

Means followed by the same lower case letter in the column and the same upper case letter in the row do not differ, by Tukey's test, at 5% significance level.

For the average speed of germination, there was no significant interaction between the factors form of extraction and concentrations. For the different concentrations, the control treatment and the concentrations 40, 80 and 100% of extract presented the same among themselves and the concentration of 20% with the highest index of average speed of germination (0.072 days).According to the results obtained for the average time of germination, there was also no interaction between the factors tested, as well as there were no significant differences for the forms of extraction. For the different concentrations, the concentration of 20 and 60% extract differed from each other, with 13.81 and 14.67 days. Corsato et al. (2010), testing aqueous extracts of sunflower (Asteraceae), verified interference on seeds of invasive plants, promoting average time of germination of black prickly pear achenes, in higher concentrations compared to the control treatment.For the average root length, there was interaction between the form of extraction and concentrations, and for the forms of extraction, the extracts differed only in the concentrations of 40 and 60% of extract, in which the extraction in crushed form presented the lowest results (2.85 and 2.92 cm). In the static extraction form, the concentrations from 20% of extract differed from the control treatment.The presented results demonstrate the dose dependent effect, as well as reports of Seung et al., (2002), in which the aqueous extract of *Tagetes minuta* leaves, inhibited the callus induction and the growth of forage turnip (*Raphanus sativus* L.), in a proportional way to the used concentrations, acting in a negative way. Similar results were described by Narwal (2002), because in tests using different concentrations of aqueous extracts of *Helianthus annuus* L., from the same family as clove, it was observed that there was a negative interference in the germination percentage, as well as a reduction in the average root length in seeds of mustard (*Brassica juncea* L.), which belongs to the same family as turnip. The present results of this experiment indicate that clove extracts interfered in a negative way mainly in the average germination percentage and in the root length of turnip seeds.According to the data presented for rubefruit seeds, for the germination percentage, there was interaction among the extraction forms and among the different extract concentrations. For the extraction obtained by grinding, the concentrations from 60 to 100% differed from the control treatment, with 28% germination, as shown in Table 3. For the static extraction, there were no significant differences between the concentrations of 60, 80 and 100%, which differed from the control treatment with 35% germination. The lowest mean germination was 23%, in the 100% concentration of extract obtained through the static extraction form. Through the results obtained by Singh et al. (2006), one can compare the allelopathic effect of plants that have a-Pinene in their chemical composition, as well as clove, which act significantly in inhibiting the germination of cultivated plants as invasive plants, such as fedegoso (*Cassia occidentalis* L.), wheat (*Triticum aestivum* L.) and amaranth (*Amaranthus viridis* L.).

Table 3- Germination percentage, mean germination speed, mean germination time, and mean root length of marigold (*Leonorus sibiricus* L.) seeds as a function of extraction form and concentration levels of aqueous extract of clove (*Tagetes patula* L.)

Germination Percentage (%)							
Extraction	Extract Concentration						
Method	0	20	40	60	80	100	Averages
Crushed	28 BC	23 aA	27 aB	25 aAB	26 aAB	24 aAB	25,5
Static	26 aD	30 bC	28 aBC	27 aABC	25 aAB	23 aA	28
AVERAGE	31,5	26,5	27,5	26	25,5	23,5	
Average Germination Speed (days)							
Crushed	0,048	0,050	0,050	0,051	0,053	0,051	0,051 a
Static	0,046	0,047	0,058	0,054	0,053	0,057	0,053 a
AVERAGE	0,047 A	0,049 A	0,054 B	0,053 B	0,053 B	0,054 B	
Average Germination Time (days)							
Crushed	20.45 aB	19.75 aAB	20.02 bB	19.44 bAB	18.58 aA	19.66 bAB	19,33
Static	20.92 aB	22.16 bB	17.10 aA	18.19 aA	19.03 bA	17.33 aA	19,65
AVERAGE	21,30	20,95	18,56	18,82	18,81	18,50	
Average Root Length (cm)							
Crushed	1,72	1,82	1,52	1,22	0,8	0,8	1,26 b
Static	1,62	1,37	1,07	1,09	0,7	0,5	1,02 a
AVERAGE	1,67 D	1,42 C	1.30 BC	1,06 B	0,75 A	0,65 A	

Means followed by the same lower case letter in the column and the same upper case letter in the row do not differ, by Tukey's test, at 5% significance level.

According to the results presented for the mean speed of germination, there was no interaction among the tested factors, in which the concentration of 20% of extract and the control treatment presented the lowest values for the mean speed of germination, differing significantly in relation to the higher concentrations of extract (40, 60, 80 and 100%). Thus, the results obtained differ from those described by Alhammadi (2008), in which extracts obtained statically from leaves of *T. minuta* did not interfere in the speed of germination of seeds of acacia (*Acacia asak*).

For the average time of germination there was interaction between the extraction forms and between the different concentrations. The extraction forms differed in all extract concentrations, presenting the longest average time in the static extraction, with an average of 22.16 days in the concentration of 20% extract.

For the average root length parameter, there was no interaction between extraction form and different concentrations. However, the static extraction interfered in the root length and differed from the triturated extraction. In the different concentrations of extracts it could be observed that the concentrations of 80 and 100% differed significantly from the lowest concentrations, thus demonstrating the dose dependent effect.

According to Chon and Nelson (2010), many plants of the Asteraceae family have allelopathic potential on invasive species, in which, through bioassays, it can be seen that as the fraction of extract used increases, there is greater interference of allelochemicals on germination and root development.

For the percentage of germination in melon seeds, there was no interaction between the extraction forms and the different concentrations. The lowest average obtained in the forms of extraction was obtained through the triturated extraction. For the different concentrations, it can be observed that the concentrations did not differ among themselves, demonstrating not to cause major damage in the germination percentage.

These results are confirmed by Batish et al. (2007), who observed that static extracts of *Tagetes minuta* interfered in the germination and growth of invasive plants, such as tiririca (*Cyperus rotundus* L.), without causing any adverse effect on the cultivated species in question, rice (*Oryza sativa* L.).For the different concentrations, it can be observed that the concentrations did not differ among themselves, interfering positively in the static extraction, increasing the percentage of melon seeds germination. For the average speed of melon seeds germination, there was no significant interaction between the different concentrations of extract and form of extraction, according to Table 4.For the variable of average time of germination, there was no significant interaction between the different concentrations of extract. For the extraction form there was a significant difference for the crushed extraction, which presented the lowest mean for the static extraction. For the average root length, there was interaction among the extraction forms and among the different extract concentrations. For the extraction forms, the static extraction (differed significantly in all concentrations from the extraction obtained from grinding.

Table 4- Germination percentage, mean germination speed, mean germination time, and mean root length of melon (*Cucumis melo* Naudin) seeds as a function of extraction and concentration levels of aqueous extract of clove (*Tagetes patula* L.)

Extraction Method	Extract Concentration						
	0	20	40	60	80	100	Averages
Germination Percentage (%)							
Crushed	83,5	86	87,5	85	86,5	85,5	85,6 a
Static	90,5	88,5	90,5	87	91,5	92	90,0 b
AVERAGE	87 A	87,2 A	89 A	86 A	89 A	88,7 A	
Average Germination Speed (days)							
Crushed	0,070	0,069	0,069	0,072	0,072	0,069	0,070 b
Static	0,060	0,061	0,061	0,060	0,059	0,066	0,061 a
AVERAGE	0,065 A	0,065 A	0,065 A	0,066 A	0,065 A	0,068 A	
Average Germination Time (days)							
Crushed	14,23	14,47	13,87	13,85	13,85	14,34	14,19 a
Static	16,53	16,39	16,59	18,85	18,85	15,85	16,43 b
AVERAGE	15,38 A	15,39 A	15,43 A	15,23 A	15,35 A	15,10 A	
Average Root Length (cm)							
Crushed	6.18 aB	6.08 bA	6.33 bAB	6.20 bAB	6.38 bAB	5.80 bA	6,27
Static	5.93 aE	5.35 toDE	4.98 aCD	4.40 aBC	4.13 aB	3.30 aA	4,68
AVERAGE	6,39	5,71	5,65	5,30	5,25	4,55	

Means followed by the same lower case letter in the column and the same upper case letter in the row do not differ, by Tukey's test, at 5% significance level.

For the crushed extraction, the 100% concentration differed from the control, but did not differ from the other concentrations. For static extraction, the 100% concentration showed a significant difference among all extract concentrations, with 3.30 cm.

The different concentrations of clove extracts presented variations from 3.30 to 6.85 cm of average root length, results that are in agreement with the reports of Scrivanti et al., (2003), in which the essential oil of *T. minuta presented* inhibitory activity on the germination and root growth of mastic (*Schinus areira* L.), paralyzing root growth for up to 96 hours, due to the presence of the allelochemicals limonene and ocimenone.

The results obtained for the percentage of germination of watermelon seeds showed that there was interaction between the extraction forms and between the different extract concentrations. For the form of extraction, there was significant difference only in the concentration of 100%, in which the static extraction presented the lowest result.

Thus, as described by Aquino and Cajazeira (2008), allelopathy is increasingly being used in the management and control of invasive plants in the cultivation of curcubits in family farming, because some species are selective in relation to allelochemical compounds, not interfering significantly in cultivated plants, because under natural conditions, it is not possible to find high concentrations, such as 100%.

For average germination speed, there was no interaction between concentrations and extraction forms. There was no significant difference between the different concentrations tested, however for the forms of extraction, the average static extraction (0.061 day) was significantly lower compared to the average ground extraction (0.066 day).

For the variable of average time of germination of watermelon seeds there was no interaction between the extraction forms and between the different concentrations. The crushed extraction form (15.25 days) differed significantly in relation to the static extraction (16.30 days), presenting the shortest average time. Thus, through the results obtained, it can be observed that watermelon seeds took longer to germinate in the static form extraction.

For the parameter of average root length of watermelon seeds, there was interaction between the extraction form and different concentrations. The static extraction form differed significantly in all extract concentrations to the crushed form extraction.

These results are confirmed by reports from Reigosa et al. (2001), who, in evaluating the effect of aqueous extracts obtained from two forms of triturated and static extraction of *Pittosporum undulatum* (frankincense) leaves on soybean seed development, found that in the static extract there was a reduction in mean root length.

For the concentrations, in the crushed extraction, the control treatment and the concentration of 100% of extracts differed from each other and from the other concentrations, with 9.10 cm and 6.45 cm of length, respectively. The extract obtained statically showed the lowest results compared to the triturated extraction, this difference may be due to the fact that the static extracts remained at rest for 24 hours, which did not happen in the triturated extraction.

However, possibly there was not enough time for the extraction of the allelochemicals, therefore, the extraction of secondary metabolites in the triturated extraction occurred in smaller proportions, thus reducing the interference in the analyzed variables, in relation to the static extraction, because according to reports by SIMÕES et al. (2004), the cold extraction methods can be influenced by the extraction time.

According to Kil and Shim (2006), similar results to the present work were obtained with static extracts of *Tagetes minuta* L. at concentrations of 10, 50 and 100% stimulated germination and average root length in gherkin (*Lotus corniculatus*) seeds.

For the seeds of aster (*Aster scaber*) and kiss of maiden (*Bidens bipinnata* L.), the extract concentrations inhibited in 72% the germination percentage, reducing the average root length in up to 80%, and even completely inhibiting root hair formation. These results confirm the hypothesis that some plants are selective in their responses to others, and may interfere with or stimulate germination and growth patterns due to their sensitivity to allelochemicals.

Conclusion

Under the conditions in which the present work was developed, the aqueous extract of clove showed allelopathic potential on the invasive plants tested, interfering negatively on them, however, it presented variable responses on the cultivated plants, depending on the concentration and extraction form.

References

ALHAMMADI, A.S.A. Allelopathic effect of *Tagetes minuta* L. water extracts on seeds germination and seedling root growth of Acacia Asak. **Assiut University Bulletin for Environmental Researches**, Assiut, v. 11, n. 1, p. 17-24, 2008.

ALVES, C.C.F. et al. Allelopathic activity of glycosylated alkaloids from *Solanum crinitum* Lam. **Floresta e Ambiente**, Rio de Janeiro, v.10, n.1, p.93-97, 2003.

AQUINO, A.R.L.; CAJAZEIRA, J.P. Manejo e controle de plantas weeds no cultivo do melão. **Embrapa Agroindústria Tropical**, Fortaleza, v. 17, p. 1-8, 2008.

BARATELLI, T. G.; **Study of plant allelopathic properties: investigation of allelochemical substances in Terminalia catappa L. (Combretaceae)**. 2006. 203 f. Dissertation (Master's) - Federal University of Rio de Janeiro, Rio de Janeiro, 2006.

BARATTO, L. et al. Investigation of allelopathic and antimicrobial activities of *Mikania laevigata* (Asteraceae) obtained from hydroponic and traditional cultivation. **Brazilian Journal of Pharmacognosy**, Curitiba. v. 18, n. 4, p. 577-582, 2008.

BATISH, D.R. et al. Crop allelopathy and its role in ecological agriculture. **Journal of Crop Production**, New Delhi, v.4, p.121-161, 2001.

BRASIL. Ministry of Agriculture, Livestock and Supply. **Regras para análises de sementes**. Brasília: MAPA/ACS, 2009. 398p.

CHAMORRO, E. R. et al. Chemical composition of essential oil from *Tagetes minuta* L. leaves and flowers. **Journal of the Argentine Chemical Society**, Buenos Aires, v. 96, n.

1-2, p. 80-86, 2008. CHON, S.U.; NELSON, C.J. Allelopathy in Compositae plants. A review. **Agronomy Sustaintable and Development**. Madison. v. 30, n. 2. p.349-358, 2010.

CORSATO, J.M.et al. Allelopathic effect of aqueous extract of sunflower leaves on the germination of soybean and black walnut. **Semina: Agrarian Sciences**, Londrina, v. 31, n. 2, p. 353-360, 2010.

FERRAZ, S.; FREITAS, L.G. **O controle de fitonematóides por plantas antagonistas e produtos naturais. Manual de Fitopatologia - princípios e conceitos**. São Paulo: Ceres, 3 ed. 1995. 782p.

FERREIRA, A.G.; AQUILA, M.E.A. Alelopatia: Uma área emergente da ecofisiologia. **Revista Brasileira de Fisiologia Vegetal**, Campinas, v.12, p.175-204, 2000.

HADAS, A. Water uptake and germination of leguminous seeds under changing external water potencial in osmotic solution. **Journal Express Botany**, Saint Louis, v. 27, p. 480-489, 1976.

HUANG, H. C.; CHOU, C. H. Impact of Plant Disease Biocontrol and Allelopathy on Biodiversity and Agricultural Sustainability. **Plant Pathology Bulletin**, Alberta, v. 14, p.1-12, 2005.

ILORI O.J. et al. Allelopatic activies of some weeds in the Asteraceae family. **International Journal of Botany**, Dusseldorf. v. 6 n. 2, p. 161-163, 2010.

KIL, B.S.; YUN, K.Y. Allelopathic *Artemisia princes* effects of water extracts of var. *orientalis* on selected plant species. **Journal of Chemical Ecology**, Iksan. v. 18, n. 1, p. 39-51, 1992.

KIL, B. S. et al. Allelopathic effects of *Artemisia lavandulaefolia*. **Korean Journal Ecology**, Iksan, v. 23, n. 2, p.149-155, 2000.

KIL, J.H. et al. Allelopathy of *Tagetes minuta* L. aqueous extracts on seed germination and root hair growth. **Korean Journal Ecology**, Iksan, v. 1, n. 3, p.171-174, 2002.

KIL, J.H.; SHIM, K.C. Allelopathic effects of *Tagetes minuta* L. and *Eupatorium rugosum* Houtt. Aqueous extracts on seedling growth of some plants. **Allelopathy Journal, Washington**, v. 18, n. 2, p. 315-322, 2006.

NARWAL, S.S. et al. Allelopathic effect of aqueous extracts of sunflower (*Helianthus annuus* L.) root on some winter oil eed crops. Geobios, **Jodhpur. v. 29, n. 4, p. 225-228, 2002.**

PEREZ, F. S. C.; LIMA. Clove (*Tagetes patula* L.) as an attractive plant for thrips (Thysanoptera) and Hymenoptera parasitoids (Hymenoptera) in protected cultivation. 2004.

REIGOSA, M.J. et al. Comparison of physiological effects of allelochemicals and comercial herbicides. **Allelopathy Journal**, v. 82, p. 211-220, 2001.

RICE, E. L. **Allelopathy**. New York: Academic Press. 1984, 353 p.

SCRIVANTI, L.R.et al. ***Tagetes minuta*** and ***Schinus areira*** essential oils as allelopathic agents. **Biochemical Systematics and Ecology**, West Chester, v. 31, p. 563-572, 2003.

SEUNG, Y.L.; KEW, C.S.; JI, H.K. Phytotoxic effect of aqueous extracts and essential oils from southern marigold (***Tagetes minuta***). **New Zealand Journal of Crop and Horticultural Science**, Thorndon, v. 30, p. 161-169, set. 2002.

SINGH, H.P. et al. a-Pinene Inhibits Growth and Induces Oxidative Stress in Roots. Annals of Botany, **Oxfor Journal**, Oxford. v. 98, p. 1261-1269, 2006.
SILVA, H.L. et al. Determination of indicator species and comparison of sunflower genotypes for allelopathic potential. **Planta Daninha**, Viçosa. v. 27 n. 4, p. 655-673, 2009.

SIMÕES, C.M.O et al. **Farmacognosia: da planta ao medicamento**. 5 ed. Porto Alegre: UFRGS, 2004. 1102p.

TAIZ, L.; ZEIGER, E. **Plant Physiology**. 4. ed. Porto Alegre: Artmed editora, 2009. 719 p.

yes
I want morebooks!

Buy your books fast and straightforward online - at one of world's fastest growing online book stores! Environmentally sound due to Print-on-Demand technologies.

Buy your books online at
www.morebooks.shop

Kaufen Sie Ihre Bücher schnell und unkompliziert online – auf einer der am schnellsten wachsenden Buchhandelsplattformen weltweit! Dank Print-On-Demand umwelt- und ressourcenschonend produzi ert.

Bücher schneller online kaufen
www.morebooks.shop

KS OmniScriptum Publishing
Brivibas gatve 197
LV-1039 Riga, Latvia
Telefax: +371 686 204 55

info@omniscriptum.com
www.omniscriptum.com

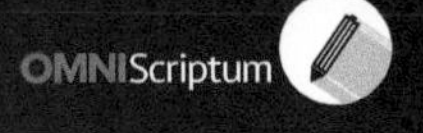

Printed by Books on Demand GmbH, Norderstedt / Germany